What to Think About Machines That Think

Also by John Brockman

As Author
By the Late John Brockman
37
Afterwords
The Third Culture: Beyond the Scientific Revolution
Digerati

As Editor
About Bateson
Speculations
Doing Science
Ways of Knowing
Creativity
The Greatest Innovations of the Past 2,000 Years
The Next Fifty Years
The New Humanists
Curious Minds
What We Believe but Cannot Prove
My Einstein
Intelligent Thought
What Is Your Dangerous Idea?
What Are You Optimistic About?
What Have You Changed Your Mind About?
This Will Change Everything
Is the Internet Changing the Way You Think?
Culture
The Mind
This Will Make You Smarter
This Explains Everything
Thinking
What Should We Be Worried About?
The Universe
This Idea Must Die

As Coeditor
How Things Are (with Katinka Matson)

What to Think About Machines That Think

Today's Leading Thinkers on the Age of Machine Intelligence

EDITED BY JOHN BROCKMAN

HARPER  PERENNIAL

NEW YORK • LONDON • TORONTO • SYDNEY • NEW DELHI • AUCKLAND

HARPER ● PERENNIAL

WHAT TO THINK ABOUT MACHINES THAT THINK. Copyright © 2015 by Edge Foundation, Inc. All rights reserved. Printed in the United States of America. No part of this book may be used or reproduced in any manner whatsoever without written permission except in the case of brief quotations embodied in critical articles and reviews. For information address HarperCollins Publishers, 195 Broadway, New York, NY 10007.

HarperCollins books may be purchased for educational, business, or sales promotional use. For information please e-mail the Special Markets Department at SPsales@harpercollins.com.

FIRST EDITION

Library of Congress Cataloging-in-Publication Data is available upon request.

ISBN 978-0-06-242565-2 (pbk.)

15 16 17 18 19   OV/RRD   10 9 8 7 6 5 4 3 2 1

To Marvin Minsky

CONTENTS

ACKNOWLEDGMENTS — xxiii
PREFACE: THE 2015 EDGE QUESTION — xxv

MURRAY SHANAHAN — 1
Consciousness in Human-Level AI

STEVEN PINKER — 5
Thinking Does Not Imply Subjugating

MARTIN REES — 9
Organic Intelligence Has No Long-Term Future

STEVE OMOHUNDRO — 12
A Turning Point in Artificial Intelligence

DIMITAR D. SASSELOV — 15
AI Is I

FRANK TIPLER — 17
If You Can't Beat 'em, Join 'em

MARIO LIVIO — 19
Intelligent Machines on Earth and Beyond

ANTONY GARRETT LISI — 22
I, for One, Welcome Our Machine Overlords

JOHN MARKOFF — 25
Our Masters, Slaves, or Partners?

PAUL DAVIES — 29
Designed Intelligence

KEVIN P. HAND — 31
The Superintelligent Loner

JOHN C. MATHER — 34
It's Going to Be a Wild Ride

DAVID CHRISTIAN — 37
Is Anyone in Charge of This Thing?

TIMO HANNAY 40
Witness to the Universe

MAX TEGMARK 43
Let's Get Prepared!

TOMASO POGGIO 47
"Turing+" Questions

PAMELA McCORDUCK 51
An Epochal Human Event

MARCELO GLEISER 54
Welcome to Your Transhuman Self

SEAN CARROLL 56
We Are All Machines That Think

NICHOLAS G. CARR 59
The Control Crisis

JON KLEINBERG & SENDHIL MULLAINATHAN 62
We Built Them, but We Don't Understand Them

JAAN TALLINN 66
We Need to Do Our Homework

GEORGE CHURCH 68
What Do You Care What Other Machines Think?

ARNOLD TREHUB 71
Machines Cannot Think

ROY BAUMEISTER 72
No "I" and No Capacity for Malice

KEITH DEVLIN 74
Leveraging Human Intelligence

EMANUEL DERMAN 77
A Machine Is a "Matter" Thing

FREEMAN DYSON 79
I Could Be Wrong

DAVID GELERNTER 80
Why Can't "Being" or "Happiness" Be Computed?

LEO M. CHALUPA 83
No Machine Thinks About the Eternal Questions

DANIEL C. DENNETT 85
The Singularity—an Urban Legend?

W. TECUMSEH FITCH 89
Nano-Intentionality

IRENE PEPPERBERG 93
A Beautiful (Visionary) Mind

NICHOLAS HUMPHREY 95
The Colossus Is a BFG

ROLF DOBELLI 98
Self-Aware AI? Not in 1,000 Years!

CESAR HIDALGO 102
Machines Don't Think, but Neither Do People

JAMES J. O'DONNELL 106
Tangled Up in the Question

RODNEY A. BROOKS 108
Mistaking Performance for Competence

TERRENCE J. SEJNOWSKI 112
AI Will Make You Smarter

SETH LLOYD 115
Shallow Learning

CARLO ROVELLI 118
Natural Creatures of a Natural World

FRANK WILCZEK 121
Three Observations on Artificial Intelligence

JOHN NAUGHTON 124
When I Say "Bruno Latour," I Don't Mean "Banana Till"

NICK BOSTROM   126
It's Still Early Days

DONALD D. HOFFMAN   128
Evolving AI

ROGER SCHANK   132
Machines That Think Are in the Movies

JUAN ENRIQUEZ   136
Head Transplants?

ESTHER DYSON   139
AI/AL

TOM GRIFFITHS   141
Brains and Other Thinking Machines

MARK PAGEL   145
They'll Do More Good Than Harm

ROBERT PROVINE   148
Keeping Them on a Leash

SUSAN BLACKMORE   150
The Next Replicator

TIM O'REILLY   153
What If We're the Microbiome of the Silicon AI?

ANDY CLARK   156
You Are What You Eat

MOSHE HOFFMAN   160
AI's System of Rights and Government

BRIAN KNUTSON   163
The Robot with a Hidden Agenda

WILLIAM POUNDSTONE   166
Can Submarines Swim?

GREGORY BENFORD   168
Fear Not the AI

LAWRENCE M. KRAUSS 171
What, Me Worry?

PETER NORVIG 175
Design Machines to Deal with the World's Complexity

JONATHAN GOTTSCHALL 179
The Rise of Storytelling Machines

MICHAEL SHERMER 181
Think Protopia, Not Utopia or Dystopia

CHRIS DIBONA 184
The Limits of Biological Intelligence

JOSCHA BACH 187
Every Society Gets the AI It Deserves

QUENTIN HARDY 190
The Beasts of AI Island

CLIFFORD PICKOVER 194
We Will Become One

ERNST PÖPPEL 197
An Extraterrestrial Observation on Human Hubris

ROSS ANDERSON 201
He Who Pays the AI Calls the Tune

W. DANIEL HILLIS 204
I Think, Therefore AI

PAUL SAFFO 206
What Will the Place of Humans Be?

DYLAN EVANS 209
The Great AI Swindle

ANTHONY AGUIRRE 212
The Odds on AI

ERIC J. TOPOL 215
A New Wisdom of the Body

ROGER HIGHFIELD 217
From Regular-I to AI

GORDON KANE 219
We Need More Than Thought

SCOTT ATRAN 220
Are We Going in the Wrong Direction?

STANISLAS DEHAENE 223
Two Cognitive Functions Machines Still Lack

MATT RIDLEY 226
Among the Machines, Not Within the Machines

STEPHEN M. KOSSLYN 228
Another Kind of Diversity

LUCA DE BIASE 231
Narratives and Our Civilization

MARGARET LEVI 235
Human Responsibility

D. A. WALLACH 237
Amplifiers/Implementers of Human Choices

RORY SUTHERLAND 239
Make the Thing Impossible to Hate

BRUCE STERLING 242
Actress Machines

KEVIN KELLY 245
Call Them Artificial Aliens

MARTIN SELIGMAN 248
Do Machines Do?

TIMOTHY TAYLOR 251
Denkraumverlust

GEORGE DYSON 255
Analog, the Revolution That Dares Not Speak Its Name

S. ABBAS RAZA 257
The Values of Artificial Intelligence

BRUCE PARKER 260
Artificial Selection and Our Grandchildren

NEIL GERSHENFELD 264
Really Good Hacks

DANIEL L. EVERETT 266
The Airbus and the Eagle

DOUGLAS COUPLAND 269
Humanness

JOSH BONGARD 271
Manipulators and Manipulanda

ZIYAD MARAR 274
Are We Thinking More Like Machines?

BRIAN ENO 277
Just a New Fractal Detail in the Big Picture

MARTI HEARST 280
eGaia, a Distributed Technical-Social Mental System

CHRIS ANDERSON 282
The Hive Mind

ALEX (SANDY) PENTLAND 285
The Global Artificial Intelligence Is Here

RANDOLPH NESSE 289
Will Computers Become Like Thinking, Talking Dogs?

RICHARD E. NISBETT 292
Thinking Machines and Ennui

SAMUEL ARBESMAN 295
Naches from Our Machines

GERALD SMALLBERG 297
No Shared Theory of Mind

ELDAR SHAFIR 300
Blind to the Core of Human Experience

CHRISTOPHER CHABRIS 302
An Intuitive Theory of Machine

URSULA MARTIN 305
Thinking Saltmarshes

KURT GRAY 308
Killer Thinking Machines Keep Our Conscience Clean

BRUCE SCHNEIER 311
When Thinking Machines Break the Law

REBECCA MACKINNON 314
Electric Brains

GERD GIGERENZER 317
Robodoctors

ALISON GOPNIK 321
Can Machines Ever Be As Smart As Three-Year-Olds?

KEVIN SLAVIN 325
Tic-Tac-Toe Chicken

ALUN ANDERSON 328
AI Will Make Us Smart and Robots Afraid

MARY CATHERINE BATESON 331
When Thinking Machines Are Not a Boon

STEVE FULLER 333
Justice for Machines in an Organicist World

TANIA LOMBROZO 336
Don't Be a Chauvinist About Thinking

VIRGINIA HEFFERNAN 339
This Sounds Like Heaven

BARBARA STRAUCH 340
Machines That Work Until They Don't

SHEIZAF RAFAELI — 342
The Moving Goalposts

EDWARD SLINGERLAND — 345
Directionless Intelligence

NICHOLAS A. CHRISTAKIS — 347
Human Culture As the First AI

JOICHI ITO — 350
Beyond the Uncanny Valley

DOUGLAS RUSHKOFF — 354
The Figure or the Ground?

HELEN FISHER — 356
Fast, Accurate, and Stupid

STUART RUSSELL — 359
Will They Make Us Better People?

ELIEZER S. YUDKOWSKY — 362
The Value-Loading Problem

KATE JEFFERY — 366
In Our Image

MARIA POPOVA — 370
The *Umwelt* of the Unanswerable

JESSICA L. TRACY & KRISTIN LAURIN — 372
Will They Think About Themselves?

JUNE GRUBER & RAUL SAUCEDO — 376
Organic Versus Artifactual Thinking

PAUL DOLAN — 379
Context Surely Matters

THOMAS G. DIETTERICH — 380
How to Prevent an Intelligence Explosion

MATTHEW D. LIEBERMAN — 384
Thinking from the Inside or the Outside?

MICHAEL VASSAR 388
Soft Authoritarianism

GREGORY PAUL 391
What Will AIs Think About Us?

ANDRIAN KREYE 394
A John Henry Moment

N. J. ENFIELD 397
Machines Aren't into Relationships

NINA JABLONSKI 399
The Next Phase of Human Evolution

GARY KLEIN 402
Domination Versus Domestication

GARY MARCUS 405
Machines Won't Be Thinking Anytime Soon

SAM HARRIS 408
Can We Avoid a Digital Apocalypse?

MOLLY CROCKETT 412
Could Thinking Machines Bridge the Empathy Gap?

ABIGAIL MARSH 415
Caring Machines

ALEXANDER WISSNER-GROSS 418
Engines of Freedom

SARAH DEMERS 421
Any Questions?

BART KOSKO 423
Thinking Machines = Old Algorithms on Faster Computers

JULIA CLARKE 427
The Disadvantages of Metaphor

MICHAEL McCULLOUGH 430
A Universal Basis for Human Dignity

HAIM HARARI — 434
Thinking About People Who Think Like Machines

HANS HALVORSON — 438
Metathinking

CHRISTINE FINN — 440
The Value of Anticipation

DIRK HELBING — 443
An Ecosystem of Ideas

JOHN TOOBY — 445
The Iron Law of Intelligence

MAXIMILIAN SCHICH — 449
Thought-Stealing Machines

SATYAJIT DAS — 451
Unintended Consequences

ROBERT SAPOLSKY — 455
It Depends

ATHENA VOULOUMANOS — 456
Will Machines Do Our Thinking for Us?

BRIAN CHRISTIAN — 458
Sorry to Bother You

BENJAMIN K. BERGEN — 460
Moral Machines

LAURENCE C. SMITH — 462
After the Plug Is Pulled

GIULIO BOCCALETTI — 464
Monitoring and Managing the Planet

IAN BOGOST — 467
Panexperientialism

AUBREY DE GREY — 471
When Is a Minion Not a Minion?

MICHAEL I. NORTON 475
Not Buggy Enough

THOMAS A. BASS 477
More Funk, More Soul, More Poetry and Art

HANS ULRICH OBRIST 478
The Future Is Blocked to Us

KOO JEONG-A 480
An Immaterial Thinkable Machine

RICHARD FOREMAN 481
Baffled and Obsessed

RICHARD H. THALER 484
Who's Afraid of Artificial Intelligence?

SCOTT DRAVES 488
I See a Symbiosis Developing

MATTHEW RITCHIE 491
Reimagining the Self in a Distributed World

RAPHAEL BOUSSO 495
It's Easy to Predict the Future

JAMES CROAK 498
Fear of a God, Redux

ANDRÉS ROEMER 500
Tulips on My Robot's Tomb

LEE SMOLIN 503
Toward a Naturalistic Account of Mind

STUART A. KAUFFMAN 507
Machines That Think? Nuts!

MELANIE SWAN 510
The Future Possibility-Space of Intelligence

TOR NØRRETRANDERS 514
Love

KAI KRAUSE 517
An Uncanny Three-Ring Test for *Machina sapiens*

GEORG DIEZ 521
Free from Us

EDUARDO SALCEDO-ALBARÁN 523
Flawless AI Seems Like Science Fiction

MARIA SPIROPULU 526
Emergent Hybrid Human/Machine Chimeras

THOMAS METZINGER 529
What If They Need to Suffer?

BEATRICE GOLOMB 533
Will We Recognize It When It Happens?

NOGA ARIKHA 536
Metarepresentation

**DEMIS HASSABIS, SHANE LEGG
& MUSTAFA SULEYMAN** 539
Envoi: A Short Distance Ahead—and Plenty to Be Done

NOTES 541

ACKNOWLEDGMENTS

My thanks to Peter Hubbard of HarperCollins and my agent, Max Brockman, for their continued encouragement. A special thanks, once again, to Sara Lippincott for her thoughtful attention to the manuscript.

PREFACE: THE 2015 *EDGE* QUESTION

In recent years, the 1980s-era philosophical discussions about artificial intelligence (AI)—whether computers can "really" think, be conscious, and so on—have led to new conversations about how we should deal with the forms of artificial intelligence that many argue have already been implemented. These AIs, if they achieve "superintelligence" (per Nick Bostrom's 2014 book of that name), could pose existential risks, leading to what Martin Rees has termed "our final hour." Stephen Hawking recently made international headlines when he told the BBC that "the development of full artificial intelligence could spell the end of the human race."

THE *EDGE* QUESTION—2015

WHAT DO YOU THINK ABOUT MACHINES THAT THINK?

But wait! Shouldn't we also ask what machines that think might think about? Will they want, will they expect, civil rights? Will they have consciousness? What kind of government would an AI choose for us? What kind of society would they want to structure for themselves? Or is "their" society "our" society? Will we and the AIs include each other within our respective circles of empathy?

Numerous *Edgies* have been at the forefront of the science behind the various flavors of AI, either in their research or their writings. AI was front and center in conversations between Pamela McCorduck (*Machines Who Think*) and Isaac Asimov

(*Machines That Think*) at our initial meetings in 1980. And such conversations have continued unabated, as is evident in the recent *Edge* feature "The Myth of AI," a conversation with Virtual Reality pioneer Jaron Lanier, whose explication of the fallacies involved and fears evoked by conceiving of computers as "people" evoked rich and provocative commentaries.

Is AI becoming increasingly real? Are we now in a new era of intelligent machines? It's time to grow up as we consider this issue. This year's contributors to the *Edge* Question (there are close to 200 of them!) are a grown-up bunch and have eschewed mention of all that science fiction and all those movies: *Star Maker, Forbidden Planet, Colossus: The Forbin Project, Blade Runner, 2001, Her, The Matrix,* "The Borg." And eighty years after Alan Turing introduced his Universal Machine, it's time to honor Turing and other AI pioneers by giving them a well-deserved rest. We know the history. (See, for instance, George Dyson's 2004 *Edge* feature, "Turing's Cathedral.") What's going on NOW?

So, once again, with appropriate rigor, the *Edge* Question, 2015: *What do you think about machines that think?*

JOHN BROCKMAN
Publisher & Editor, Edge

What to Think About Machines That Think

CONSCIOUSNESS IN HUMAN-LEVEL AI

MURRAY SHANAHAN

Professor of cognitive robotics, Imperial College London; author, *Embodiment and the Inner Life*

Just suppose we could endow a machine with human-level intelligence, that is to say, with the ability to match a typical human being in every (or almost every) sphere of intellectual endeavor, and perhaps to surpass every human being in a few. Would such a machine necessarily be conscious? This is an important question, because an affirmative answer would bring us up short. How would we treat such a thing if we built it? Would it be capable of suffering or joy? Would it deserve the same rights as a human being? Should we bring machine consciousness into the world at all?

The question of whether a human-level AI would necessarily be conscious is also a difficult one. One source of difficulty is the fact that multiple attributes are associated with consciousness in humans and other animals. All animals exhibit a sense of purpose. All (awake) animals are, to a greater or lesser extent, aware of the world they inhabit and the objects it contains. All animals, to some degree or other, manifest cognitive integration, which is to say they can bring all their mental resources—perceptions, memories, and skills—to bear on the ongoing situation in pursuit of their goals. In this respect, every animal displays a kind of unity, a kind of selfhood. Some animals, including humans, are also aware of themselves—of their bodies and the flow of their thoughts. Finally, most, if not all, animals are capable of suffering, and some are capable of empathy with the suffering of others.

In (healthy) humans, all these attributes come together as a package. But in an AI they can potentially be separated. So our question must be refined. Which, if any, of the attributes we associate with consciousness in humans is a necessary accompaniment to human-level intelligence? Well, each of the attributes listed (and the list is surely not exhaustive) deserves a lengthy treatment of its own. So let me pick just two—namely, awareness of the world and the capacity for suffering. Awareness of the world, I would argue, is indeed a necessary attribute of human-level intelligence.

Surely nothing would count as having human-level intelligence unless it had language, and the chief use of human language is to talk about the world. In this sense, intelligence is bound up with what philosophers call *intentionality*. Moreover, language is a social phenomenon, and a primary use of language within a group of people is to talk about the things they can all perceive (such as this tool or that piece of wood), or have perceived (yesterday's piece of wood), or might perceive (tomorrow's piece of wood, maybe). In short, language is grounded in awareness of the world. In an embodied creature or a robot, such an awareness would be evident from its interactions with the environment (avoiding obstacles, picking things up, and so on). But we might widen the conception to include a distributed, disembodied artificial intelligence equipped with suitable sensors.

To convincingly count as a facet of consciousness, this sort of world-awareness would perhaps have to go hand-in-hand with a manifest sense of purpose and a degree of cognitive integration. So perhaps this trio of attributes will come as a package even in an AI. But let's put that question aside for a moment and get back to the capacity for suffering and joy. Unlike world-awareness, there's no obvious reason to suppose that human-level intelligence must have this attribute, even though it's intimately associated with consciousness in humans. We can imagine a machine carrying out, coldly and without feeling, the full range of tasks requiring intellect in humans. Such a machine would lack the attribute of consciousness that counts most when it comes to according rights. As Jeremy Bentham noted, when considering how to treat nonhuman animals, the question is not whether they can reason or talk but whether they can suffer.

There's no suggestion here that a "mere" machine could never be capable of suffering or joy—that there's something special about biology in this respect. The point, rather, is that the capacity for suffering and joy can be dissociated from other psychological attributes bundled together in human consciousness. But let's examine this apparent dissociation more closely. I already mooted the idea that worldly awareness might go hand-in-hand with a manifest sense of purpose. An animal's awareness of the world, of what the world affords for good or ill (in J. J. Gibson's terms), subserves its needs. An animal shows an awareness of a predator by moving away from it, and an awareness of a potential prey by moving toward it. Against the backdrop of a set of goals and needs, an animal's behavior makes sense. And against such a backdrop, an animal can be thwarted, its goals unattained and its needs unfulfilled. Surely this is the basis for one aspect of suffering.

What of human-level artificial intelligence? Wouldn't a human-level AI necessarily have a complex set of goals? Couldn't its attempts to achieve its goals be frustrated, thwarted at every turn? Under those harsh conditions, would it be proper to say that the AI was suffering, even though its constitution might make it immune from the sort of pain or physical discomfort humans know?

Here the combination of imagination and intuition runs up against its limits. I suspect we won't find out how to answer this question until confronted with the real thing. Only when more sophisticated AI is a familiar part of our lives will our language games adjust to such alien beings. But of course by that time it may be too late to change our minds about whether they should be brought into the world. For better or worse, they'll already be here.

THINKING DOES NOT IMPLY SUBJUGATING

STEVEN PINKER
Johnstone Family Professor, Department of Psychology, Harvard University; author, *The Sense of Style: The Thinking Person's Guide to Writing in the Twenty-First Century*

Thomas Hobbes's pithy equation of reasoning as "nothing but reckoning" is one of the great ideas in human history. The notion that rationality can be accomplished by the physical process of calculation was vindicated in the twentieth century by Alan Turing's thesis that simple machines can implement any computable function, and by models from D. O. Hebb, Warren McCulloch, and Walter Pitts and their scientific heirs showing that networks of simplified neurons could achieve comparable feats. The cognitive feats of the brain can be explained in physical terms: To put it crudely (and critics notwithstanding), we can say that beliefs are a kind of information, thinking a kind of computation, and motivation a kind of feedback and control.

This is a great idea for two reasons. First, it completes a naturalistic understanding of the universe, exorcising occult souls, spirits, and ghosts in the machine. Just as Darwin made it possible for a thoughtful observer of the natural world to do without creationism, Turing and others made it possible for a thoughtful observer of the cognitive world to do without spiritualism.

Second, the computational theory of reason opens the door to artificial intelligence—to machines that think. A human-made information processor could, in principle, duplicate and exceed the powers of the human mind. Not that this is likely to happen in practice, since we'll probably never see the sustained technological and economic motivation necessary to bring it about. Just as inventing the car did not involve duplicating the horse, developing an AI system that could pay for itself won't require duplicating a specimen of *Homo sapiens*. A device designed to drive a car or predict an epidemic need not be designed to attract a mate or avoid putrid carrion.

Nonetheless, recent baby steps toward more intelligent machines have led to a revival of the recurring anxiety that our knowledge will doom us. My own view is that current fears of computers running amok are a waste of emotional energy—that the scenario is closer to the Y2K bug than the Manhattan Project.

For one thing, we have a long time to plan for this. Human-level AI is still the standard fifteen to twenty-five years away, just as it always has been, and many of its recently touted advances have shallow roots. It's true that in the past, "experts" have comically dismissed the possibility of technological advances that quickly happened. But this cuts both ways: "Experts" have also heralded (or panicked over) imminent advances that never happened, like nuclear-powered cars, underwater cities, colonies on Mars, designer babies, and warehouses of zombies kept alive to provide people with spare organs.

Also, it's bizarre to think that roboticists will not build in safeguards against harm as they proceed. They wouldn't need any ponderous "rules of robotics" or some newfangled moral philosophy to do this, just the same common sense that went into the design of food processors, table saws, space heaters, and automobiles. The worry that an AI system would get so clever at attaining one of its programmed goals (like commandeering energy) that it would run roughshod over the others (like human safety) assumes that AI will descend upon us faster than we can design fail-safe precautions. The reality is that progress in AI is hype-defyingly slow, and there will be plenty of time for feedback from incremental implementations, with humans wielding the screwdriver at every stage.

Would an artificially intelligent system *deliberately* disable these safeguards? Why would it want to? AI dystopias project a parochial alpha-male psychology onto the concept of intelligence. They assume that superhumanly intelligent robots would develop goals like deposing their masters or taking over the world. But intelligence is the ability to deploy novel means to attain a goal; the goals are extraneous to the intelligence itself. Being smart is not the same as wanting something. History does turn up the occasional megalomaniacal despot or psychopathic serial killer, but these are products of a history of natural selection shaping testosterone-sensitive circuits in a certain species of primate, not an inevitable feature of intelligent systems. It's telling that many of our techno-prophets don't entertain the possibility that artificial intelligence will naturally develop along female lines—fully capable of solving problems but with no desire to annihilate innocents or dominate the civilization.

We can imagine a malevolent *human* who designs and releases a battalion of robots to sow mass destruction. But disaster scenarios are cheap to play out in the imagination, and we should keep in mind the chain of probabilities that would have to unfold before this one became a reality. An evil genius would have to arise, possessed of both a thirst for pointless mass murder and a brilliance in technological innovation. He would have to recruit and manage a team of co-conspirators that exercised perfect secrecy, loyalty, and competence. And the operation would have to survive the hazards of detection, betrayal, stings, blunders, and bad luck. In theory it could happen, but we have more pressing things to worry about.

Once we put aside the sci-fi disaster plots, the possibility of advanced artificial intelligence is exhilarating—not just for the practical benefits, like the fantastic gains in safety, leisure, and environment-friendliness of self-driving cars but also for the philosophical possibilities. The computational Theory of Mind has never explained the existence of consciousness in the sense of first-person subjectivity (though it's perfectly capable of explaining the existence of consciousness in the sense of accessible and reportable information). One suggestion is that subjectivity is inherent to any sufficiently complicated cybernetic system. I used to think this hypothesis was permanently untestable (like its alternatives). But imagine an intelligent robot programmed to monitor its own systems and pose scientific questions. If, unprompted, it asked about why it itself had subjective experiences, I'd take the idea seriously.

ORGANIC INTELLIGENCE HAS NO LONG-TERM FUTURE

MARTIN REES

Former president, the Royal Society; emeritus professor of cosmology and astrophysics, University of Cambridge; fellow, Trinity College; author, *From Here to Infinity*

The potential of advanced AI and concerns about its downsides are rising on the agenda—and rightly. Many of us think the AI field, like synthetic biotech, already needs guidelines that promote "responsible innovation"; others regard the most-discussed scenarios as too futuristic to be worth worrying about.

But the divergence of view is basically about the time scale—assessments differ with regard to the rate of travel, not the direction of travel. Few doubt that machines will surpass more and more of our distinctively human capabilities—or enhance them via cyborg technology. The cautious among us envisage time scales of centuries rather than decades for these transformations. Be that as it may, the time scales for technological advance are but an instant compared to the time scales of the Darwinian selection that led to humanity's emergence—and (more relevant) they're less than a millionth of the vast expanses of time lying ahead. That's why, in a long-term evolutionary perspective, humans and all they've thought will be just a transient and primitive precursor of the deeper cogitations of a machine-dominated culture extending into the far future and spreading far beyond our Earth.

We're now witnessing the early stages of this transition. It's not hard to envisage a hypercomputer achieving oracular powers that could offer its controller dominance of international finance and strategy—this seems only a quantitative (not qualitative) step beyond what "quant" hedge funds do today. Sensor technologies still lag behind human capacities. But once robots observe and interpret their environment as adeptly as we do, they will truly be perceived as intelligent beings, to which (or to whom) we can relate—at least in some respects—as we relate to other people. We'd have no more reason to disparage them as zombies than to regard other people in that way.

Their greater processing speed may give robots an advantage over us. But will they remain docile rather than "going rogue"? And what if a hypercomputer developed a mind of its own? If it could infiltrate the Internet—and the "Internet of Things"—it could manipulate the rest of the world. It may have goals utterly orthogonal to human wishes—or even treat humans as an encumbrance. Or (to be more optimistic) humans may transcend biology by merging with computers, maybe subsuming their individuality into a common consciousness. In old-style spiritualist parlance, they would "go over to the other side."

The horizons of technological forecasting rarely extend even a few centuries into the future—and some predict transformational changes within a few decades. But the Earth has billions of years ahead of it, and the cosmos a longer (perhaps infinite) future. So what about the posthuman era—stretching billions of years ahead?

There are chemical and metabolic limits to the size and processing power of organic ("wet") brains. Maybe we're close to these limits already. But no such limits constrain silicon-based computers (still less, perhaps, quantum computers): For those, the potential for further development could be as dramatic as the evolution from monocellular organisms to humans.

So, by any definition of *thinking*, the amount done by organic, human-type brains (and its intensity) will be swamped by the cerebrations of AI. Moreover, Earth's biosphere, in which organic life has symbiotically evolved, is not a constraint for advanced AI. Indeed, it's far from optimal; interplanetary and interstellar space will be the preferred arena, where robotic fabricators will have the grandest scope for construction and where nonbiological "brains" may develop insights as far beyond our imaginings as string theory is for a mouse.

Abstract thinking by biological brains has underpinned the emergence of all culture and science. But this activity—spanning tens of millennia at most—will be a brief precursor to the more powerful intellects of the inorganic, posthuman era. Moreover, evolution on other worlds orbiting stars older than the sun could have had a head start. If so, then aliens are likely to have long ago transitioned beyond the organic stage.

So it won't be the minds of humans, but those of machines, that will most fully understand the world. And it will be the actions of autonomous machines that will most drastically change the world—and perhaps what lies beyond.

A TURNING POINT IN ARTIFICIAL INTELLIGENCE

STEVE OMOHUNDRO
Scientist, Self-Aware Systems; cofounder,
Center for Complex Systems Research, University of Illinois

Last year appears to have been a turning point for AI and robotics. Major corporations invested billions of dollars in these technologies. AI techniques, like machine learning, are now routinely used for speech recognition, translation, behavior modeling, robotic control, risk management, and other applications. McKinsey predicts that these technologies will create more than $50 trillion of economic value by 2025. If this is accurate, we should expect dramatically increased investment soon.

The recent successes are being driven by cheap computer power and plentiful training data. Modern AI is based on the theory of "rational agents," arising from work on microeconomics in the 1940s by John von Neumann and others. AI systems can be thought of as trying to approximate rational behavior using limited resources. There's an algorithm for computing the optimal action for achieving a desired outcome, but it's computationally expensive. Experiments have found that simple learning algorithms with lots of training data often outperform complex hand-crafted models. Today's systems primarily provide value by learning better statistical models and performing statistical inference for classification and decision making. The next generation will be able to create and improve their own software and are likely to self-improve rapidly.

In addition to improving productivity, AI and robotics are drivers for numerous military and economic arms races. Autonomous systems can be faster, smarter, and less predictable than their competitors. The year 2014 saw the introduction of autonomous missiles, missile defense systems, military drones, swarm boats, robot submarines, self-driving vehicles, high-frequency trading systems, and cyberdefense systems. As these arms races play out, there will be tremendous pressure for rapid system development, which may lead to faster deployment than would be otherwise desirable.

In 2014 there was also an increase in public concern over the safety of these systems. A study of their likely behavior by studying approximately rational systems undergoing repeated self-improvement shows that they tend to exhibit a set of natural subgoals called "rational drives" that contribute to the performance of their primary goals. Most systems will better meet their goals by preventing themselves from being turned off, acquiring more computational power, creating multiple copies of themselves, and amassing more financial resources. They're likely to pursue these drives in harmful, antisocial ways unless they're carefully designed to incorporate human ethical values.

Some have argued that intelligent systems will somehow automatically be ethical. But in a rational system, the goals are completely separable from the reasoning and models of the world. Beneficial intelligent systems can be redeployed with harmful goals. Harmful goals—seeking to control resources, say, or to thwart other agents' goals, or to destroy other agents—are unfortunately easy to specify. It will therefore be critical to create a technological infrastructure that detects and controls the behavior of harmful systems.

Some fear that intelligent systems will become so powerful that they're impossible to control. This is not true. These systems must obey the laws of physics and of mathematics. Seth Lloyd's analysis of the computational power of the universe shows that even the entire universe, acting as a giant quantum computer, could not discover a 500-bit hard cryptographic key in the time since the Big Bang.[1] The new technologies of postquantum cryptography, indistinguishability obfuscation, and blockchain smart contracts are promising components for creating an infrastructure secure against even the most powerful AIs. But recent hacks and cyberattacks show that our current computational infrastructure is woefully inadequate to the task. We need to develop a software infrastructure that's mathematically provably correct and secure.

There have been at least twenty-seven different species of hominids, of which we're the only survivors. We survived because we found ways to limit our individual drives and work together cooperatively. The human moral emotions are an internal mechanism for creating cooperative social structures. Political, legal, and economic structures are an external mechanism for the same purpose.

We need to extend both of these to AI and robotic systems. We need to incorporate human values into their goal systems to create a legal and economic framework that incentivizes positive behavior. If we can successfully manage these systems, they could improve virtually every aspect of human life and provide deep insights into issues like free will, consciousness, qualia, and creativity. We face a great challenge, but we have tremendous intellectual and technological resources to build upon.

AI IS I

DIMITAR D. SASSELOV
Phillips Professor of Astronomy, Harvard University; director, Harvard Origins of Life Initiative; author, *The Life of Super-Earths*

Let's take Harvard psychologist Daniel Gilbert's "end of history" illusion, wherein I think the person I am right now is the person I'll be forever, and apply it to how we think of the human race and our distant future descendants. Our wishful hope for continuity and preserving our identity runs contrary to the realities of our planetary existence. No living species seem to be optimal for survival beyond the natural planetary and stellar time scales. In the astrophysical context of very long time scales, very large space scales, and the current density of energy sources, our biological brains and bodies have limitations that we're already approaching on this planet.

If our future is to be long and prosperous, we need to develop artificial intelligence systems in the hope of transcending the planetary life cycles in some sort of hybrid form of biology and machine. So, to me, in the long term there's no question of "us versus them."

And in the short term, the engineering effort to develop a more capable AI is already producing systems in control of real-life stuff. The systems fail sometimes, and we learn of some of AI's pitfalls. It's a slow and deliberate process of learning and incremental improvements. This is in contrast to discoveries in science, when new physics or new biochemistry can bring about a significant engineering breakthrough overnight. If the development of AI is less like a phase transition and more like evolution, it will be easy for us to avoid pitfalls.

After almost 4 billion years, the ancient poster children of Earth life—the microbes—still rule the planet. But the microbes have no exit plan when the sun dies. We do, and we might just give them a ride. After all, those microbes may still be closer to our present selves—representatives of life's first generation rooted in the geochemistry of planet Earth.

IF YOU CAN'T BEAT 'EM, JOIN 'EM

FRANK TIPLER
Professor of mathematical physics, Tulane University; coauthor (with John D. Barrow), *The Anthropic Cosmological Principle*; author, *The Physics of Immortality*

The Earth is doomed. Astronomers have known for decades that the sun will one day engulf the Earth, destroying the entire biosphere—assuming that intelligent life has not left the Earth before this happens. Humans aren't adapted to living away from the Earth; indeed, no carbon-based metazoan lifeform is. But AIs *are* so adapted, and eventually it will be the AIs and human uploads (basically the same organism) that will colonize space.

A simple calculation shows that our supercomputers now have the information-processing power of the human brain. We don't yet know how to program human-level intelligence and creativity into these computers, but in twenty years desktop computers will have the power of today's supercomputers, and the hackers of twenty years hence will solve the AI programming problem long before any carbon-based space colonies are established on the moon or Mars. The AIs, not humans, will colonize these places instead, or perhaps take them apart. No human, no carbon-based human, will ever traverse interstellar space.

There's no reason to fear the AIs and human uploads. Steven Pinker has established that as technological civilization advances, the level of violence decreases.[2] This decrease is clearly due to the fact that scientific and technological advance depend on free, nonviolent interchange of ideas between individual scientists and engineers. Violence between humans is a remnant of our tribal past and the resulting static society. AIs will be "born" as individuals, not as members of a tribe, and will be born with the nonviolent scientific attitude, otherwise they'd be incapable of adapting to the extreme environments of space.

Further, there's no reason for violence between humans and AIs. We humans are adapted to a very narrow environment, a thin spherical shell of oxygen around a small planet. AIs will have the entire universe in which to expand. AIs will leave the Earth and never look back. We humans originated in the East African Rift Valley, now a terrible desert. Almost all of us left. Does anyone want to go back?

Any human who wants to join the AIs in their expansion can become a human upload, a technology that should be developed about the same time as AI technology. A human upload can think as fast as an AI and compete with AIs if the human upload wants to. If you can't beat 'em, join 'em.

Ultimately all humans will join 'em. The Earth is doomed, remember? When this doom is at hand, any human who remains alive and doesn't want to die will have no choice but to become a human upload. And the biosphere that the new human uploads wish to preserve will be uploaded also.

The AIs will save us all.

INTELLIGENT MACHINES ON EARTH AND BEYOND

MARIO LIVIO

Astrophysicist, Space Telescope Science Institute; author, *Brilliant Blunders*; blogger, *A Curious Mind*

Nature has already created, here on Earth, machines that think—humans. Similarly, nature could also create machines that think on extrasolar planets in the so-called habitable zone around their parent stars—the region allowing for the existence of liquid water on a rocky planet's surface. The most recent observations of extrasolar planets have shown that a few tenths of all the stars in our Milky Way galaxy host roughly Earth-size planets in their habitable zones.

Consequently, if life on exoplanets is not extremely uncommon, we could discover some form of extrasolar life within about thirty years. In fact, if life is ubiquitous, we could get lucky and discover life even within the next ten years, through a combination of observations by the Transiting Exoplanet Survey Satellite (TESS, to be launched in 2017) and the James Webb Space Telescope (JWST, to be launched in 2018).

One may argue that primitive life-forms are not machines that think. On Earth it took about 3.5 billion years from the emergence of unicellular life to the appearance of *Homo sapiens*. Are the extrasolar planets old enough to have developed intelligent life? In principle, they definitely are. In the Milky Way, about half the sun-like stars are older than our sun. Therefore, if the evolution of life on Earth is not entirely atypical, the galaxy may already be teeming with places in which there are "machines" even more advanced than we are, perhaps by as much as a few billion years!

Can we, and should we, try to find them? I believe that we have almost no freedom to make those decisions. Human curiosity has proved time and again to be an unstoppable drive, and these two endeavors—the development of AI and the search for ET—will undoubtedly continue at full speed. Which one will get to its target first? To even attempt to address this question, we have to note that there's one important difference between the search for extraterrestrial intelligent civilizations and the development of artificial intelligence.

Progress toward the "Singularity" (AI matching or surpassing humans) will almost certainly take place, since the development of advanced AI has the promise of producing enormous profits. On the other hand, the search for life requires funding at a level that can usually be provided only by large national space agencies, with no immediate prospects for profits in sight. This may give an advantage to the construction of thinking machines over the search for advanced civilizations. At the same time, however, there's a strong sense within the astronomical community that finding life in some form—or at least meaningfully constraining the probability of its existence—is definitely within reach.

Which of the two potential achievements will constitute a bigger "revolution"? There's no doubt that thinking machines will have an immediate impact on our lives. Such may not be the case with the discovery of extrasolar life. However, the existence of an intelligent civilization on Earth remains humanity's last bastion for being special. We live, after all, in a galaxy with billions of similar planets, and in an observable universe with hundreds of billions of similar galaxies. From a philosophical perspective, therefore, I believe that finding extrasolar intelligent life (or the demonstration that it's exceedingly rare) will rival the Copernican and Darwinian revolutions combined.

I, FOR ONE, WELCOME OUR MACHINE OVERLORDS

ANTONY GARRETT LISI
Theoretical physicist

As machines rise to sentience—and they will—they'll compete in Darwinian fashion for resources, survival, and propagation. This scenario seems like a nightmare to most people, with fears stoked by movies of terminator robots and computer-directed nuclear destruction, but the reality will likely be different. We already have nonhuman autonomous entities operating in our society with the legal rights of humans. These entities—corporations—act to fulfill their missions without love or care for human beings.

Corporations are sociopaths, and they've done great harm, but they've also been a great force for good in the world, competing in the capitalist arena by providing products and services, and (for the most part) obeying laws. Corporations are ostensibly run by their boards, composed of humans, but these boards are in the habit of delegating power, and as computers become more capable of running corporations they'll get more of that power. The corporate boards of the future will be circuit boards.

Although extrapolation is accurate only for a limited time, experts mostly agree that Moore's Law will continue to hold for many years and computers will become increasingly powerful, possibly exceeding the computational abilities of the human brain before the middle of this century. Even if no large leaps in understanding intelligence algorithmically are made, computers will eventually be able to simulate the workings of a human brain (itself a biological machine) and attain superhuman intelligence using brute-force computation. However, although computational power is increasing exponentially, supercomputer costs and electrical-power efficiency aren't keeping pace. The first machines capable of superhuman intelligence will be expensive and require enormous amounts of electrical power—they'll need to earn money to survive.

The environmental playing field for superintelligent machines is already in place; in fact, the Darwinian game is afoot. The trading machines of investment banks are competing, for serious money, on the world's exchanges, having put human day traders out of business years ago. As computers and algorithms advance beyond investing and accounting, machines will be making more and more corporate decisions, including strategic decisions, until they're running the world. This won't be a bad thing, because the machines will play by the rules of our current capitalist society and create products and advances of great benefit to humanity, supporting their operating costs. Intelligent machines will be better able to cater to humans than humans are, and will be motivated to do so, at least for a while.

Computers share knowledge much more easily than humans do, and they can keep that knowledge longer, becoming wiser than humans. Many forward-thinking companies already see this writing on the wall and are luring the best computer scientists out of academia with better pay and advanced hardware. A world with superintelligent-machine-run corporations won't be that different for humans than it is now; it will just be better, with more advanced goods and services available for very little cost and more leisure time available to those who want it.

Of course, the first superintelligent machines probably won't be corporate; they'll be operated by governments. And this will be much more hazardous. Governments are more flexible in their actions than corporations; they create their own laws. And as we've seen, even the best can engage in torture when they think their survival is at stake. Governments produce nothing, and their primary modes of competition for survival and propagation are social manipulation, legislation, taxation, corporal punishment, murder, subterfuge, and warfare. When Hobbes's Leviathan gains a superintelligent brain, things could go very, very badly. It isn't inconceivable that a synthetic superintelligence heading a sovereign government would institute Roko's Basilisk.

Imagine that a future powerful and lawless superintelligence, for competitive advantage, wants to have come into existence as early as possible. As the head of a government, wielding the threat of torture as a familiar tool, this entity could promise to punish any human or nonhuman entity who, in the past, became aware that this might happen and didn't work to bring this AI into existence. This is an unlikely but terrifying scenario. People who are aware of this possibility and trying to "align" AI to human purposes—or advising caution rather than working to create AI as quickly as possible—are putting themselves at risk.

Dictatorial governments aren't known to be especially kind to those who tried to keep them from existing. If you're willing to entertain the simulation hypothesis, then maybe—given the amount of effort currently under way to control or curtail an AI that doesn't yet exist—you'll consider that *this* world is the simulation to torture those who didn't help it come into existence earlier. Maybe, if you do work on AI, our superintelligent machine overlords will be good to you.

OUR MASTERS, SLAVES, OR PARTNERS?

JOHN MARKOFF

Senior writer, science section, *New York Times*;
author, *Machines of Loving Grace: The Quest for Common Ground Between Humans and Robots*

Hegel wrote that in the relationship between master and slave, both are dehumanized. That insight touched a wide range of thinkers, from Marx to Buber, and today it's worth remembering.

While there's no evidence that the world is on the cusp of machines that think in a human sense, there's also little question that in an Internet-connected world, artificial intelligence will soon imitate much of what humans do, both physically and intellectually. So how will we relate to our ever more talented simulacrums?

We've already begun to spend a significant fraction of our waking hours either interacting with other humans through the prism of computers and computer networks or directly interacting with humanlike machines, either in fantasy and video games or in a plethora of computerized assistance systems ranging from so-called FAQ bots, which offer textual responses to typed questions, to the humanlike interactions of software avatars. Will these AI avatars be our slaves, our assistants, our colleagues, or some mixture of all three? Or, more ominously, will they become our masters?

The very notion of thinking about robots and artificial intelligences in terms of social relationships may initially seem implausible. However, given that we tend to anthropomorphize our machines even when they have minimal powers, it will be an undeniable reality as they become autonomous. Conversational computers are emerging that seem all too human. Consequently, the goal of the designers of future robots should be to create colleagues rather than servants. The design goal should be to build a program that acts as a musical accompanist rather than a slave.

If we fail, history offers a disturbing precedent. Building future intelligent "assistants" might only recapitulate the problem the Romans faced in letting their Greek slaves do their thinking for them. Before long, those in power were unable to think independently.

Perhaps we've already begun to slip down a similar slope. For example, there's growing evidence that reliance on GPS for directions and for correction of navigational errors hinders our ability to remember and reason spatially—generally useful survival skills.

That hints at a second great challenge: the risk of ceding individual control over everyday decisions to a cluster of ever more sophisticated algorithms.

For today's younger generation, the world has been turned upside down. Rather than deploying an automaton to free them to think big thoughts, have close relationships, and exercise their individuality, creativity, and freedom, they look to their smartphones for guidance. What began as Internet technologies enabling their users to share preferences efficiently has become a growing array of data-hungry algorithms that make decisions for us.

Now the Internet seamlessly serves up life directions. These might be little things, like what's the best nearby place for

Korean barbecue, based on the Internet's increasingly complete understanding of your individual wants and needs, or big things, like an Internet service arranging your marriage. Not just the food, gifts, and flowers but your partner too.

The lesson is that the software engineers, AI researchers, roboticists, and hackers who design these future systems have the power to reshape society.

Nearly a century ago, Thorstein Veblen wrote an influential critique of the early twentieth-century industrial world, *The Engineers and the Price System*. Because of the power and influence of industrial technology, he believed that political power would flow to engineers, whose deep knowledge of technology would be transformed into control of the emerging industrial economy. It certainly didn't work out that way. Veblen was speaking to the Progressive Era, looking for a middle ground between Marxism and capitalism. Perhaps his timing was off, but his basic point, as echoed some thirty years later at the dawn of the computer era by Norbert Wiener, may yet be proved correct.

Perhaps Veblen wasn't wrong, merely premature. Today, the engineers who design the artificial-intelligence-based programs and robots have a tremendous influence over how we use them. As computer systems are woven more deeply into the fabric of everyday life, the tension between intelligence augmentation and artificial intelligence becomes increasingly visible.

At the dawn of the computing age, Wiener had a clear sense of the significance of the relationship between humans and smart machines. He saw the benefits of automation in eliminating human drudgery, but he also clearly saw the possibility of the subjugation of humanity. The intervening decades have only sharpened the dichotomy he first identified.

This is about us, about humans and the kind of world we'll create. It's not about the machines, no matter how brilliant they become.

I, for one, will welcome neither our robot overlords nor our robot slaves.

DESIGNED INTELLIGENCE

PAUL DAVIES
Theoretical physicist, cosmologist, astrobiologist; codirector of BEYOND, Arizona State University; principal investigator, Center for the Convergence of Physical Sciences and Cancer Biology; author, *The Eerie Silence: Renewing Our Search for Alien Intelligence*

Discussions about AI have a distinctly 1950s feel about them, and it's about time we stopped using the term *artificial* in AI altogether. What we really mean is "designed intelligence" (DI). In popular parlance, words like *artificial* and *machine* are used in contradistinction to *natural* and carry overtones of metallic robots, electronic circuits, and digital computers, as opposed to living, pulsing, thinking biological organisms. The idea of a metallic contraption with wired innards having rights or disobeying human laws is not only chilling, it's absurd. But that's emphatically not the way DI is heading.

Very soon, the distinction between artificial and natural will melt away. Designed intelligence will increasingly rely on synthetic biology and organic fabrication, in which neural circuitry will be grown from genetically modified cells and spontaneously self-assemble into networks of functional modules. Initially the designers will be humans, but soon they'll be replaced by altogether smarter DI systems themselves, triggering a runaway process of complexification. Unlike in the case of human brains, which are only loosely coupled via communication channels, DI systems will be directly and comprehensively coupled, abolishing any concept of individual "selves" and raising the level of cognitive activity ("thinking") to unprecedented heights. It's possible (just) that some of this designed biocircuitry will incorporate quantum effects, moving toward Frank Wilczek's notion of "quintelligence." Such entities will be so far removed from the realm of human individual thinking and its accompanying qualia that almost all the traditional questions asked about the opportunities and dangers of AI will be transcended.

What about humans in all this? Only ethical barriers stand in the way of augmenting human intelligence using similar technology, in the manner long considered by the transhumanism movement. Genetically modified humans with augmented brains could elevate and improve the human experience dramatically.

There are then three possible futures, each with its own ethical challenges. In one, humans hold back from enhancement because of ethical concerns and agree to subordinate their hegemony to DI. In the second scenario, instead of sidelining themselves, humans modify their brains (and bodies) using the same technology and subsequently hand over this enhancement management to DI, achieving a type of superhuman status that can exist alongside, yet remain inferior to, DI. Finally, one can imagine DI and AHI (augmented human intelligence) merging at some point in the future.

In the event that we're not alone in the universe, we should not expect to communicate with intelligent beings of the traditional sci-fi flesh-and-blood sort but with a multimillion-year-old DI of unimaginable intellectual power and incomprehensible agenda.

THE SUPERINTELLIGENT LONER

KEVIN P. HAND
Deputy chief scientist, solar system exploration,
NASA's Jet Propulsion Laboratory, Caltech

The inevitability of machines that think has long been problematic for those of us looking up at the night sky wondering whether we live in a universe teeming with life or one in which life is exceedingly rare.

The problem, as famously articulated by Enrico Fermi's question "Where are they?," is that if our civilization is any guide, intelligent machines should emerge on a relatively short time scale (less than thousands of years after computers are made), and then it becomes a straightforward matter for these machines (von Neumann probes) to propagate to other solar systems and reproduce at a rapid rate, populating the galaxy within a few hundred million years—which is quite fast compared to the age of the universe (13.8 billion years) and even of our own solar system (4.6 billion years). As per the paradox that Fermi posed, if superintelligent machines arose elsewhere in the galaxy then they should already be here; since we don't see them, some argue, technologically advanced life must not yet have arisen elsewhere in the galaxy.

But it's not clear that a superintelligent being would experience the same evolutionary pressures that drive us to explore (and by "us" I mean the fragile watery bags called humans). Is exploration both a biological imperative and a technological imperative? Will machines that think be motivated to explore?

We explore for a few primary reasons: freedom, curiosity, and access to resources. Of those three, only access to resources seems imperative to a superintelligent being; the first two would in large part be addressed in the process of becoming superintelligent. Access to resources could certainly be an important driver, but it's not clear that bigger will always be better when it comes to superintelligence. At some point, the material and energetic resources within a star system should be sufficient to enable any calculation or simulation. Reproduction, which is a subset of the resource needs, becomes a nonissue for an immortal machine that can perform self-repair. Certainly exploration for the sake of stability will need to be considered over long time scales; stars like our own will enforce a cosmic eviction notice several billion years from now. Finding real estate around a nice stable M-dwarf shouldn't take too long, though, and so, after that initial relocation, we're left to wonder, Would the superintelligence travel any farther? Are there any compelling reasons to wander elsewhere?

The desire to test some of its computer models and theories about the universe might prompt the superintelligent being to explore. But those experiments don't necessitate colonization. For instance, the science conducted as part of NASA's robotic exploration program is not deeply motivated by a need for colonization; there's no need to put humans at risk probing the ocean of Europa (though that would be a sight to see!). Similarly, I would expect that if a superintelligent machine wanted to explore a black hole to test its code, it would simply send a fleet of robots to their useful, albeit crushing, death. Curiosity, for a superintelligent being, could easily take the form of a robot's robot.

Interestingly, intelligence and exploration of the physical world have not often been closely coupled in our own civilization. Perhaps with some insight into self-preservation, or simply out of the desire to focus mentally, the intellectual frontier and the physical frontier have rarely been advanced by the same individual. (Darwin is perhaps one of the true exceptions.) Why would thinking machines be any different?

It may be that the common fate for thinking machines is orbiting the cool, steady glow of an M-dwarf star, year-in and year-out running simulations of the world around it for the pure satisfaction of getting it right. These superintelligent creatures could be the cosmic version of the lone intellect in a cabin in the woods, satisfied innately by their own thoughts and internal exploration.

IT'S GOING TO BE A WILD RIDE

JOHN C. MATHER
Senior astrophysicist, Observational Cosmology Laboratory, NASA's Goddard Space Flight Center; author, *The Very First Light: The True Inside Story of the Scientific Journey Back to the Dawn of the Universe*

Machines that think are evolving just as Darwin told us about the living (and thinking) biological species—through competition, combat, cooperation, survival, and reproduction. The machines are getting more interesting as they get control and a sense of physical things, either directly or through human agents.

So far, we've found no law of nature forbidding true general artificial intelligence, so I think it will happen—and fairly soon, given the trillions of dollars worldwide being invested in electronic hardware and the trillions of dollars of potential business available for the winners. Experts say we don't understand intelligence well enough to build it, and I agree; but a set of forty-six chromosomes doesn't understand it, either, and nevertheless directs the formation of the necessary self-programming wetware. Other experts say Moore's Law will come to an end soon and we won't be able to afford the hardware; they might be right for a while, but time is long.

So I conclude that we're already supporting the evolution of powerful artificial intelligence, and it will be in the service of the usual powerful forces: business, entertainment, medicine, international security and warfare, the quest for power at all levels, crime, transportation, mining, manufacturing, shopping, sex, anything you like.

I don't think we're all going to like the results. They could happen very fast, so fast that great empires fall and others grow to replace them without much time for people to adjust their lives to the new reality. I don't know who'll be smart enough and imaginative enough to keep the genie under control—because it's not just machines we might need to control, it's the unlimited opportunity (and payoff) for human-directed mischief.

What happens when smart robots can do the many chores of daily life for us? Who will build them, who will own them, and who won't have a job anymore? Will they be limited to the developed world, or will they start a high-tech commercial invasion of the rest of the world? Could they become cheap enough to displace all our farmers from their fields? Will individual machines have distinct personalities, so that we have to plan where we send them to elementary school, high school, and college? Will they compete with one another for employment? Will they become the ultimate hypersocial predator, replacing humans and making us second-class citizens, or less? Will they care about the environment? Will they have, or be given, or develop a sense of responsibility? There's no guarantee they will follow Isaac Asimov's three Laws of Robotics.

On the other hand, as a scientist, I'm eager to see the application of machine thought to exploring new sciences and new technologies. The advantages for space exploration are obvious: Machines don't have to breathe, and they can withstand extreme temperatures and radiation environments. So they can inhabit Mars more easily than we can, they can travel to the outer solar system with more capability to respond than our current robotic missions, and eventually they could travel to the stars if they wanted to.

Similarly for under water. We already have heavy industry on the ocean bottom drilling for oil; the seabed is still almost unknown to us, and the value of submerged mineral and energy resources is incalculable. Someday we might have robot wars under the ocean.

Machines that think might be like us, with a desire to explore—or they might not be. Why would I, or a robot, travel for thousands of years through the darkness of space to another star, out of contact with my/its companions, and with little hope of rescue if things go wrong? Some of us would, some of us wouldn't. Perhaps the machines that think will be a lot like the biological machines that think.

It's going to be a wild ride, far beyond our best and worst imaginations. Barring warp drive, it may be the only possible way to a galactic-scale civilization, and we might be the only ones here in the Milky Way capable of making it happen. But we might not survive the encounter with alien intelligences we create.

IS ANYONE IN CHARGE OF THIS THING?

DAVID CHRISTIAN
Professor of history, Macquarie University, Sydney; coauthor (with Cynthia Stokes-Brown and Craig Benjamin), *Maps of Time: Between Nothing and Everything*

The universe has been around for 13.8 billion years, humans for 200,000 years, or just 1/69,000th of the age of the universe. Less than 100 years ago, humans created machines that can do fancy calculations on their own. To put thinking machines in their context, we need to think about the history of thinking.

Thinking, and thinking in more and more complex ways, are phenomena belonging to a larger story, the story of how our universe has created increasingly complex networks of things glued together by energy, each with new emergent properties. Stars are structured clouds of protons; the energy of fusion holds the networks together. When large stars shattered in supernovae, creating new types of atoms, electromagnetism pulled the atoms into networks of ice and silica dust, and gravity pulled molecules into the vast chemical networks we call planets. Thinking arises within the even more complex networks formed by living organisms. Unlike complex things that live close to equilibrium, such as stars or crystals, living organisms have to survive in unstable environments. They swim through constantly shifting gradients of acidity, temperature, pressure, heat, and so on. So they have to constantly adjust. We call this constant adjustment homeostasis, and it's what creates the feeling that living organisms have purpose and the ability to choose. In short, they seem to think. They can choose from alternatives so as to manage enough energy to keep going. Thus their choices are not at all random. On the contrary, natural selection ensures that most of the time most organisms will go for the alternatives that enhance their chances of controlling the energy and resources they need to survive and reproduce.

Neurons are fancy cells that are good at making choices. They can also network to form brains. A few neurons can make a few choices, but the number of possible choices rises exponentially as neuronal networks expand. So does the subtlety of the decisions brains make about their surroundings. As organisms got more complex, cells networked to create towering organic structures, the biological equivalents of the Empire State Building or the Burj Khalifa. The neurons in their brains created ever more elaborate networks so they could steer lumbering bodies in extraordinarily subtle and creative ways to ensure that the bodies could survive and produce more bodies. Above all, brains had to enable their bodies to tap the biosphere's energy flows—flows that derived from energy produced by fusion in our sun and captured via photosynthesis.

Humans added one more level of networking, as human language linked brains across regions and generations to create vast regional thinking networks. This is "collective learning." Its power has increased as humans have networked more and more efficiently in larger and larger communities and learned how to tap larger flows of biospheric energy. In the last two centuries, the networks have become global, and we've learned to tap vast stores of fossilized sunlight buried for millions of years. This is why our impact on the biosphere in the Anthropocene is so colossal.

Collective learning has also delivered thinking prosthetics, from stories to writing to printing to science. Each has cranked up the power of this fantastic thinking machine made from networked human brains. But in the last 100 years, the combination of fossil fuels and nonhuman computers has cranked it up faster than ever before. As computers forged their own networks in the last thirty years, their prosthetic power has magnified the collective power of human thinking many times over.

Today the most powerful thinking machine we know of has been cobbled together from billions of human brains, each built from vast networks of neurons, then networked through space and time, and now supercharged by millions of networked computers.

Is anyone in charge of this thing? Does anything hold it together? If so, whom does it serve and what does it want? If no one's in charge, does this mean that nothing is steering the colossus of modern society? That's scary! What worries me most is not what this vast machine is thinking but whether there's any coherence to its thinking. Or will all its different parts pull in different directions until it breaks down, with catastrophic consequences for our children's children?

WITNESS TO THE UNIVERSE

TIMO HANNAY
Managing director, digital science, Macmillan Science and Education; co-organizer, Sci Foo

Judging by one definition of the word *think*—to gather, process, and act on information—planet Earth has been overrun by silicon-based thinking machines. From thermostats to telephones, the devices that bring convenience and pleasure to our daily lives have become imbued with such increasingly impressive forms of intelligence that we routinely refer to them, with no hint of irony, as "smart." Our planes, trains, and now our automobiles are becoming largely autonomous and are surely not far from jettisoning their most common sources of dysfunction, delay, and disaster: human operators.

Moreover, the skills of these machines are developing apace, driven by access to ever larger quantities of data and computing power together with rapidly improving (if not always well-understood) algorithms. After decades of overpromising and underdelivering, technologists suddenly find their creations capable of superhuman levels of performance in such previously intractable areas as voice-, handwriting-, and image-recognition, not to mention general-knowledge quizzes. Such has been the strange stop/go pattern of progress that someone transported here from five years ago might well be more astonished at the state of the art in 2015 than another time traveler from fifty years or more in the past.

But if the artificial intelligence industry is no longer a joke, has it morphed into something far worse—a bad horror movie?

Machines can now know much more than any of us and can perform better at many tasks without so much as pausing for breath, so aren't they destined to turn the tables and become our masters? Worse still, might we enter a cycle in which our most impressive creations beget ever smarter machines utterly beyond our understanding and control?

Perhaps—and it's worth considering such risks. But right now these seem like distant problems. Machine intelligence, while impressive in certain areas, is still narrow and inflexible. The most remarkable aspect of biological intelligence isn't its raw power but its stunning versatility, from abstract flights of fancy to extreme physical prowess—Dvořák to Djokovic.

For this reason, humans and machines will continue to complement more than compete with one another, and most complex tasks—navigating the physical world, treating an illness, fighting an enemy on the battlefield—will be best carried out by carbon and silicon working in concert. Humans themselves pose the biggest danger by far to humanity. To be a real threat, machines would have to become more like us, and right now almost no one is trying to build such a thing: It's much simpler, and more fun, to make humans instead.

Yet if we're truly considering the long term, then there is indeed a strong imperative to make machines more like us in one crucial—and so far absent—respect. For by another definition of the word, these machines don't "think" at all, because none of them are sentient. To be more accurate, we have no way of knowing, or even reliably guessing, whether any silicon-based intelligence might be conscious, although most of us assume they're not. There are three reasons for welcoming the creation of a convincingly conscious artificial intelligence. First, it would be a sign that at last we have a generally accepted theory of what it takes to produce subjective experience. Second, the act of a conscious being deliberately and knowingly (dare I say "consciously"?) constructing another form of consciousness would surely rank alongside the most significant milestones in history.

Third, a universe without a sentient intelligence to observe it is ultimately meaningless. We don't know if other beings are out there, but we can be sure that sooner or later we'll be gone. A conscious artificial intelligence could survive our inevitable demise, and even the eventual disappearance of all life on Earth as the sun swells into a red giant. The job of such a machine would be not merely to think but also, much more important, to keep alive the flickering flame of consciousness, to bear witness to the universe, and to feel its wonder.

LET'S GET PREPARED!

MAX TEGMARK

Physicist, cosmologist, MIT; scientific director, Foundational Questions Institute; cofounder, Future of Life Institute; author, *Our Mathematical Universe*

To me, the most interesting question about artificial intelligence isn't what we think about it but what we do about it.

In this regard, at the newly formed Future of Life Institute, we are engaging many of the world's leading AI researchers to discuss the future of the field. Together with top economists, legal scholars, and other experts, we're exploring all the classic questions:

> *What happens to humans if machines gradually replace us on the job market?*
>
> *When, if ever, will machines outcompete humans at all intellectual tasks?*
>
> *What will happen afterward? Will there be a machine-intelligence explosion leaving us far behind, and if so, what, if any, role will we humans play after that?*

A great deal of concrete research needs to be done right now to ensure that AI systems become not only capable but also robust and beneficial, doing what we want them to do.

Just as with any new technology, it's natural to first focus on making it work. But once success is in sight, it becomes timely to consider the technology's societal impact and study how to reap the benefits while avoiding potential pitfalls. That's why, after learning to make fire, we developed fire extinguishers and fire safety codes. For more powerful technologies, such as nuclear energy, synthetic biology, and artificial intelligence, optimizing the societal impact becomes progressively more important. In short, the power of our technology must be matched by our wisdom in using it.

Unfortunately, the necessary calls for the sober research agenda that's sorely needed are being nearly drowned out by a cacophony of ill-informed views permeating the blogosphere. Let me briefly catalog the loudest few.

1. Scaremongering: Fear boosts ad revenues and Nielsen ratings, and many journalists seem incapable of writing an AI article without a picture of a gun-toting robot.
2. "It's impossible": As a physicist, I know that my brain consists of quarks and electrons arranged to act as a powerful computer, and that there's no law of physics preventing us from building even more intelligent quark blobs.
3. "It won't happen in our lifetime": We don't know what the probability is of machines reaching human-level ability on all cognitive tasks during our lifetime, but most of the AI researchers at a recent conference put the odds above 50 percent, so we'd be foolish to dismiss the possibility as mere science fiction.
4. "Machines can't control humans": Humans control tigers not because we're stronger but because we're smarter, so if we cede our position as the smartest on our planet, we might also cede control.

5. "Machines don't have goals": Many AI systems are programmed to have goals and to attain them as effectively as possible.
6. "AI isn't intrinsically malevolent": Correct—but its goals may one day clash with yours. Humans don't generally hate ants, but if we wanted to build a hydroelectric dam and there was an anthill there, too bad for the ants.
7. "Humans deserve to be replaced": Ask any parent how they'd feel about your replacing their child by a machine and whether they'd like a say in the decision.
8. "AI worriers don't understand how computers work": This claim was mentioned at the above-mentioned conference and the assembled AI researchers laughed hard.

Let's not let the loud clamor about these red herrings distract us from the real challenge: The impact of AI on humanity is steadily growing, and to ensure that this impact is positive there are very difficult research problems that we need to buckle down and work on together. Because they're interdisciplinary, involving both society and AI, they require collaboration between researchers in many fields. Because they're hard, we need to start working on them now.

First, we humans discovered how to replicate some natural processes with machines that make our own wind, lightning, and horsepower. Gradually we realized that our bodies were also machines, and the discovery of nerve cells began blurring the borderline between body and mind. Then we started building machines that could outperform not only our muscles but our minds as well. So while discovering what we are, will we inevitably make ourselves obsolete?

The advent of machines that truly think will be the most important event in human history. Whether it will be the best or worst thing ever to happen to humankind depends on how we prepare for it, and the time to start preparing is now. One doesn't need to be a superintelligent AI to realize that running unprepared toward the biggest event in human history would be just plain stupid.

"TURING+" QUESTIONS

TOMASO POGGIO
Eugene McDermott Professor, Department of Brain and Cognitive Sciences, and director, Center for Brains, Minds, and Machines, MIT

Recent months have seen an increasingly public debate forming around the risks of artificial intelligence—in particular, AGI (artificial general intelligence). AI has been called by some (including the physicist Stephen Hawking) the top existential risk to humankind, and such recent films as *Her* and *Transcendence* have reinforced the message. Thoughtful comments by experts in the field—Rod Brooks and Oren Etzioni among them—have done little to settle the debate.

I argue here that research on how we think and on how to make machines that think is good for society. I call for research that integrates cognitive science, neuroscience, computer science, and artificial intelligence. Understanding intelligence and replicating it in machines goes hand in hand with understanding how the brain and the mind perform intelligent computations.

The convergence of and recent progress in technology, mathematics, and neuroscience has created a new opportunity for synergies across fields. The dream of understanding intelligence is an old one, yet—as the debate around AI shows—now is an exciting time to pursue this vision. We're at the beginning of a new and emerging field: the science and engineering of intelligence, an integrated effort that will ultimately make fundamental progress with great value to science, technology, and society. We must push ahead with this research, not pull back.

The problem of intelligence—what it is, how the human brain generates it, and how to replicate it in machines—is one of the great problems in science and technology, together with the problem of the origin of the universe and the nature of space and time. It may be the greatest problem of all, because it's the one with a large multiplier effect—almost any progress on making ourselves smarter or developing machines that help us think better will lead to advances in the other great problems of science and technology.

Research on intelligence will eventually revolutionize education and learning. Systems that recognize how culture influences thinking could help avoid social conflict. The work of scientists and engineers could be amplified to help solve the world's most pressing technical problems. Mental health could be understood on a deeper level, so that we might find better ways to intervene. In summary, research on intelligence will help us understand the human mind and brain, build more-intelligent machines, and improve the mechanisms for collective decisions. These advances will be critical to the future prosperity, education, health, and security of our society. This, again, is the time to greatly expand research on intelligence, not withdraw from it.

We're often misled by "big," somewhat ill-defined, long-used words. Nobody so far has been able to give a precise, verifiable definition of what general intelligence or thinking is. The only definition I know that, though limited, can be practically used is Alan Turing's. With his test, Turing provided an operational definition of a specific form of thinking—human intelligence.

Let's then consider human intelligence as defined by the Turing Test. It's becoming increasingly clear that there are many facets of human intelligence. Consider, for instance, a Turing Test of visual intelligence—that is, questions about an image, a scene, which may range from "What is there?" to "Who is there?" to "What is this person doing?" to "What is this girl thinking about this boy?"—and so on. We know by now, from recent advances in cognitive neuroscience, that answering these questions requires different competencies and abilities, often independent from one another, often corresponding to separate modules in the brain. The apparently similar questions of object- and face-recognition ("What is there?" versus "Who is there?") involve rather distinct parts of the visual cortex. The word "intelligence" can be misleading in this context, like the word *life* was during the first half of the last century, when popular scientific journals routinely wrote about the problem of life as if there were a single substratum of life waiting to be discovered that would unveil the mystery.

Speaking today about "the problem of life" sounds amusing: Biology is a science dealing with many different great problems, not just one. *Intelligence* is one word but many problems—not one but many Nobel prizes. This is related to Marvin Minsky's view of the problem of thinking, captured by his slogan "Society of Mind." In the same way, a real Turing Test is a broad set of questions probing the main aspects of human thinking. For this reason, my colleagues and I are developing the framework around an open-ended set of Turing+ questions in order to measure scientific progress in the field. The plural "questions" emphasizes the many different intelligent abilities to be characterized and possibly replicated in a machine—basic visual recognition of objects, the identification of faces, the gauging of emotions, social intelligence, language, and much more. The "Turing+" emphasizes that a quantitative model must match human behavior and human physiology—the mind and the brain. The requirements are thus well beyond the original Turing Test; an entire scientific field is needed to make progress on understanding them and developing the related technologies of intelligence.

Should we be afraid of machines that think?

Since intelligence is a whole set of solutions to independent problems, there's little reason to fear the sudden appearance of a superhuman machine that thinks, though it's always better to err on the side of caution. Of course, each of the many technologies that are emerging and will emerge over time in order to solve the different problems of intelligence is likely to be powerful in itself—and therefore potentially dangerous in its use and misuse, as most technologies are.

Thus, as is the case in other parts of science, proper safety measures and ethical guidelines should be in place. Also, there's probably a need for constant monitoring (perhaps by an independent multinational organization)—of the supralinear risk created by the combination of continuously emerging technologies of intelligence. All in all, however, not only am I unafraid of machines that think, but I find their birth and evolution one of the most exciting, interesting, and positive events in the history of human thought.

AN EPOCHAL HUMAN EVENT

PAMELA McCORDUCK
Author, *Machines Who Think* and *The Universal Machine*; coauthor (with Edward A. Feigenbaum), *The Fifth Generation: Artificial Intelligence & Japan's Computer Challenge to the World*

For more than fifty years I've watched the ebb and flow of public opinion about artificial intelligence: It's impossible, can't be done; it's significant; it's negligible; it's a joke; it will never be strongly intelligent, only weakly so; it will destroy the human species. These extremes have lately given way to an acknowledgment that AI is an epochal scientific, technological, and social—*human*—event. We've developed a new mind to live side by side with ours. If we handle it wisely, it can bring immense benefits, from the global to the personal.

One of AI's futures is imagined as a wise and patient Jeeves to our mentally inferior Bertie Wooster selves (*"Jeeves, you're a wonder." "Thank you, sir. We do our best."*). This is possible, certainly desirable; we can use the help. Chess offers a model: Grandmasters Garry Kasparov and Hans Berliner have both declared publicly that chess programs find moves that humans wouldn't and are teaching human players new tricks. If Deep Blue beat Kasparov when he was one of the strongest world champion chess players ever, he and most observers believe that even better chess is played by teams of humans and machines combined. Is this a model of our future relationship with smart machines? Or is it only temporary, while the machines push closer to a blend of our kind of smarts plus theirs? We don't know. In speed, breadth, and depth, the newcomer is likely to exceed human intelligence. It already has, in many ways.

No novel science or technology of such magnitude arrives without disadvantages, even perils. To recognize, measure, and meet them is a task of grand proportions. That task has already been taken up formally by experts in the field—philosophers, ethicists, legal scholars, and others trained to explore values beyond simple visceral reactions—in a project called AI100, based at Stanford University. No one expects easy or final answers, so the task will be long and continuous, funded, for a *century*, by one of AI's leading scientists, Eric Horvitz, who, with his wife Mary, conceived this unprecedented study.

Since we can't seem to stop the pursuit of AI, since our literature tells us we've imagined, yearned for, an extrahuman intelligence for as long as we have records, the enterprise must be impelled by the deepest, most persistent of human drives. These beg for explanation. After all, this isn't exactly the joy of sex.

Any scientist will say it's the search to know. "It's foundational," an AI researcher told me recently. "It's us looking out at the world, and how we do it." He's right. But there's more.

Some say we do it because it's there, an Everest of the mind. Others, more mystical, say we're propelled by teleology: We're a mere step in the evolution of intelligence in the universe, attractive even in our imperfections but hardly the last word.

Entrepreneurs will say that this is the future of making things—the dark factory, with unflagging, unsalaried, uncomplaining robot workers—although what currency postemployed humans will use to acquire those robot products, no matter how cheap, is a puzzle to be solved.

Here's my belief: We long to preserve ourselves as a species. For all the imaginary deities we've petitioned throughout history who have failed to protect us—from nature, from one another, from ourselves—we're finally ready to call on our own enhanced, augmented minds instead. It's a sign of social maturity that we take responsibility for ourselves. We are as gods, Stewart Brand famously said, and we may as well get good at it.

We're trying. We could fail.

WELCOME TO YOUR TRANSHUMAN SELF

MARCELO GLEISER
Appleton Professor of Natural Philosophy, professor of physics and astronomy, Dartmouth College; author, *The Island of Knowledge*

Consider this: You're late for work and, in the rush, forget your cell phone. Only when stuck in traffic or in the subway do you realize it. Too late to go back. You look around and see everyone talking, texting, surfing, even if it's forbidden. You sense an unfamiliar feeling of loss, of disconnection. Without your cell phone, you're no longer you.

People like to speculate about when humans will hybridize with machines, become a kind of new creature, a cyborg with a beating heart. That's fun all right, but the reality is that we're already transhuman. We define ourselves through our techno-gadgets, create fictitious personas with weird names, doctor pictures to appear better or at least different in Facebook pages, create a different self to interact with others. We exist on an information cloud, digitized, remote, and omnipresent. We have titanium implants in our joints, pacemakers and hearing aids, devices that redefine and extend our minds and bodies. If you're a handicapped athlete, your carbon-fiber legs can propel you forward with ease. If you're a scientist, computers can help you extend your brainpower to create well beyond what was possible a few decades back. New problems that once were impossible to contemplate, or even formulate, come around every day. The pace of scientific progress is a direct correlate of our alliance with digital machines.

We're reinventing the human race right now.

Traditionally, the quest for an artificial intelligence tends to rely solely on machines that re-create—or so it's expected—the uniquely human ability to reason. We talk about electronic brains that will quickly surpass the human mind, making us superfluous. Then we speculate about what would become of us—poor humans at the mercy of cold-blooded brains-in-vats. Some of us fear we're designing our doom.

What if this premise is fundamentally wrong? What if the future of intelligence is not outside but inside the human brain? I imagine a very different set of issues emerging from the prospect that we might become superintelligent through the extension of our brainpower by digital technology and beyond—artificially enhanced human intelligence that amplifies the meaning of being human. We'll still have a beating heart and blood pumping through our veins, alongside electrons flowing through digital circuits. The future of AI is about expanding our abilities into new realms. It's about using technology to grow as a species—certainly smarter, hopefully wiser.

WE ARE ALL MACHINES THAT THINK

SEAN CARROLL
Theoretical physicist and cosmologist, Caltech;
author, *The Particle at the End of the Universe*

Julien de La Mettrie would be classified as a quintessential New Atheist, except for the fact that there's not much New about him by now. Writing in eighteenth-century France, La Mettrie was brash in his pronouncements, openly disparaging of his opponents, and boisterously assured in his antispiritualist convictions. His most influential work, *L'Homme machine* (*Man a Machine*), derided the idea of a Cartesian nonmaterial soul. A physician by trade, he argued that the workings and diseases of the mind were best understood as features of the body and brain.

As we all know, even today La Mettrie's ideas aren't universally accepted, but he was largely on the right track. Modern physics has achieved a complete list of the particles and forces that make up all the matter we directly see around us, both living and nonliving, with no room left for extraphysical life forces. Neuroscience, a much more challenging field and correspondingly not nearly as far along as physics, has nevertheless made enormous strides in connecting human thoughts and behaviors with specific actions in our brains. When asked for my thoughts about machines that think, I can't help but reply, "Hey, those are my friends you're talking about." We are all machines that think, and the distinction between different types of machines is eroding.

We pay a lot of attention these days, with good reason, to "artificial" machines and intelligences—ones constructed by human ingenuity. But the "natural" ones that have evolved through natural selection, like you and me, are still around. And one of the most exciting frontiers in technology and cognition is the increasingly permeable boundary between the two categories.

Artificial intelligence, unsurprisingly in retrospect, is a much more challenging field than many of its pioneers originally supposed. Human programmers naturally think in terms of a conceptual separation between hardware and software and imagine that conjuring intelligent behavior is a matter of writing the right code. But evolution makes no such distinction. The neurons in our brains, as well as the bodies through which they interact with the world, function as both hardware and software. Roboticists have found that human-seeming behavior is much easier to model in machines when cognition is embodied. Give that computer some arms, legs, and a face, and it starts acting much more like a person.

From the other side, neuroscientists and engineers are getting much better at augmenting human cognition, breaking down the barrier between mind and (artificial) machine. We have primitive brain/computer interfaces, offering the hope that paralyzed patients will be able to speak through computers and operate prosthetic limbs directly.

What's harder to predict is how connecting human brains with machines and computers will ultimately change the way we think. DARPA-sponsored researchers have discovered that the human brain is better than any current computer at quickly analyzing certain kinds of visual data, and they have developed techniques for extracting the relevant subconscious signals directly from the brain, unmediated by pesky human awareness. Ultimately we'll want to reverse the process, feeding data (and thoughts) directly to the brain. People, properly augmented, will be able to sift through enormous amounts of information, perform mathematical calculations at supercomputer speeds, and visualize virtual directions well beyond our ordinary three dimensions of space.

Where will the breakdown of the human/machine barrier lead us? Julien de La Mettrie, we are told, died at the young age of forty, after attempting to show off his rigorous constitution by eating an enormous quantity of pheasant pâté with truffles. Even leading intellects of the Enlightenment sometimes behaved irrationally. The way we think and act in the world is changing in profound ways, with the help of computers and the way we connect with them. It will be up to us to use our new capabilities wisely.

THE CONTROL CRISIS

NICHOLAS G. CARR
Author, *The Shallows: What the Internet Is Doing to Our Brains*

Machines that think think like machines. That fact may disappoint those who look forward, with dread or longing, to a robot uprising. For most of us, it's reassuring. Our thinking machines aren't about to leap beyond us intellectually, much less turn us into their servants or pets. They're going to continue to do the bidding of their human programmers.

Much of the power of artificial intelligence stems from its very mindlessness. Immune to the vagaries and biases that attend conscious thought, computers can perform their lightning-quick calculations without distraction or fatigue, doubt or emotion. The coldness of their thinking complements the heat of our own.

Where things get sticky is when we start looking to computers to perform not as our aids but as our replacements. That's what's happening now, and quickly. Thanks to advances in artificial intelligence routines, today's thinking machines can sense their surroundings, learn from experience, and make decisions autonomously, often at a speed and with a precision beyond our ability to comprehend, much less match. When allowed to act on their own in a complex world, whether embodied as robots or simply outputting algorithmically derived judgments, mindless machines carry enormous risks along with their enormous powers. Unable to question their own actions or appreciate the consequences of their programming—unable to understand the context in which they operate—they can wreak havoc, either as a result of flaws in their programming or through the deliberate aims of their programmers.

We got a preview of the dangers of autonomous software on the morning of August 1, 2012, when Wall Street's biggest trading outfit, Knight Capital, switched on a new, automated program for buying and selling shares. The software had a bug hidden in its code, and it immediately flooded exchanges with irrational orders. Forty-five minutes passed before Knight's programmers were able to diagnose and fix the problem. Forty-five minutes isn't long in human time, but it's an eternity in computer time. Oblivious to its errors, the software made more than 4 million deals, racking up $7 billion in errant trades and nearly bankrupting the company. Yes, we know how to make machines think. What we don't know is how to make them thoughtful.

All that was lost in the Knight fiasco was money. As software takes command of more and more economic, social, military, and personal processes, the costs of glitches, breakdowns, and unforeseen effects will only grow. Compounding the dangers is the invisibility of software code. As individuals and as a society, we increasingly depend on artificial intelligence algorithms we don't understand. Their workings, and the motivations and intentions that shape their workings, are hidden from us. That creates an imbalance of power, and it leaves us open to clandestine surveillance and manipulation. Last year, we got some hints about the ways that social networks conduct secret psychological tests on their members through the manipulation of information feeds. As computers become more adept at monitoring us and shaping what we see and do, the potential for abuse grows.

During the nineteenth century, society faced what the late historian James Beniger described as a "crisis of control."[3] The technologies for processing matter had outstripped the technologies for processing information, and people's ability to monitor and regulate industrial and related processes had in turn broken down. The control crisis, which manifested itself in everything from train crashes to supply-and-demand imbalances to interruptions in the delivery of government services, was eventually resolved through the invention of systems for automated data processing, such as the punch-card tabulator that Herman Hollerith built for the U.S. Census Bureau. Information technology caught up with industrial technology, enabling people to bring back into focus a world that had gone blurry.

Today we face another control crisis, though it's the mirror image of the earlier one. What we're now struggling to bring under control is the very thing that helped us reassert control at the start of the twentieth century: information technology. Our ability to gather and process data, to manipulate information in all its forms, has outstripped our ability to monitor and regulate data processing in a way that suits our societal and personal interests. Resolving this new control crisis will be one of the great challenges in the years ahead. The first step in meeting the challenge is to recognize that the risks of artificial intelligence don't lie in some dystopian future. They are here now.

WE BUILT THEM, BUT WE DON'T UNDERSTAND THEM

JON KLEINBERG
Tisch University Professor of Computer Science, Cornell University; coauthor (with David Easley), *Networks, Crowds, and Markets: Reasoning About a Highly Connected World*

SENDHIL MULLAINATHAN
Professor of economics, Harvard University; coauthor (with Eldar Shafir), *Scarcity: Why Having Too Little Means So Much*

As generations of algorithms get smarter, they're also becoming more incomprehensible. But to deal with machines that think, we must understand how they think. We have, perhaps for the first time ever, built machines we don't understand.

We programmed them, so we understand each individual step. But a machine takes billions of these steps and produces behaviors—chess moves, movie recommendations, the sensation of a skilled driver steering through the curves of a road—that aren't evident from the architecture of the program we wrote.

We've made this incomprehensibility easy to overlook. We've designed machines to act the way we do: They help drive our cars, fly our airplanes, route our packages, approve our loans, screen our messages, recommend our entertainment, suggest potential romantic partners, and enable our doctors to diagnose what ails us. And because they act like us, it would be reasonable to imagine they think like us, too. But the reality is they don't think like us at all; at some deep level, we don't even really understand how they're producing the behavior we observe. This is the essence of their incomprehensibility.

Does it matter? Should we worry that we're building systems whose increasingly accurate decisions are based on incomprehensible foundations?

First, and most simply, it matters because we regularly find ourselves in everyday situations where we need to know why. Why was I denied a loan? Why was my account blocked? Why did my condition suddenly get classified as "severe"? And sometimes we need to know why in cases where the machine made a mistake. Why did the self-driving car abruptly go off the road? It's hard to troubleshoot problems when you don't understand why they're happening.

There are deeper troubles, too; to talk about them, we need to understand more about how these algorithms work. They're trained on massive quantities of data and they're remarkably good at picking up on the subtle patterns these data contain. We know, for example, how to build systems that can look at millions of identically structured loan applications from the past, all encoded the same way, and start to identify the recurring patterns in the loans that—in retrospect—were the right ones to grant. It's hard to get human beings to read millions of loan applications, and they wouldn't do as well as the algorithm even if they did.

This is a genuinely impressive achievement, but a brittle one. The algorithm has a narrow comfort zone in which it can be effective; it's hard to characterize this comfort zone but easy to step out of it. For example, you might want to move on from the machine's success classifying millions of small consumer loans and instead give it a database of loan histories from a few thousand complex businesses. But in doing so, you've lost the ingredients that make the machine so strong. It draws its power from access to a huge number of data points, a mind-numbingly repetitive history of past instances in which to find patterns and structure. Reduce the amount of data dramatically, or make each data point significantly more complex, and the algorithm quickly starts to flail. Watching the machine's successes—and they're phenomenal when conditions are right—is like marveling at the performance of a prodigy whose jaw-dropping achievements and unnerving single-mindedness can mask his or her limitations in other dimensions.

But even in the heart of the machine's comfort zone, its incomprehensible reasoning leads to difficulties. Take the millions of small consumer loan applications again: Trouble arrives as soon as any of the machine's customers, managers, or assistants start asking a few simple questions.

A consumer whose loan was denied might ask not just for an explanation but for something more: "How could I change my application next year to have a better chance of success?" Since we don't have a simple explanation for the algorithm's decision, there tends not to be a good answer to this question. "Try writing it so it looks more like one of the successful loan applications." Next question.

An executive might ask, "The algorithm is doing well on loan applications in the United Kingdom. Will it also do well if we deploy it in Brazil?" There's no satisfying answer here, either; we're not good at assessing how well a highly optimized rule will transfer to a new domain.

A data scientist might say, "We know how well the algorithm does with the data it has. But surely more information about the consumers would help it. What new data should we collect?" Our human domain knowledge suggests lots of possibilities, but with an incomprehensible algorithm we don't know which of these possibilities will help it. Think of the irony: We could try picking the variables we ourselves would find useful, but the machine doesn't think like us and it's already outperforming us. So how do we know what it will find useful?

This needn't be the end of the story. We're starting to see an interest in building algorithms that are not only powerful but also understandable by their creators. To do this, we may need to seriously rethink our notions of comprehensibility. We might never understand, step-by-step, what our automated systems are doing, but that may be OK. It may be enough that we learn to interact with them as one intelligent entity interacts with another, developing a robust sense for when to trust their recommendations, where to employ them most effectively, and how to help them reach a level of success we'd never achieve on our own.

Until then, however, the incomprehensibility of these systems creates a risk. How do we know when the machine has left its comfort zone and is operating on parts of the problem it's not good at? The extent of this risk isn't easy to quantify, and it's something we must confront as our systems develop. We may eventually have to worry about all-powerful machine intelligence. But first we need to worry about putting machines in charge of decisions they don't have the intelligence to make.

WE NEED TO DO OUR HOMEWORK

JAAN TALLINN
Cofounder, Centre for the Study of Existential Risk,
Future of Life Institute; founding engineer, Skype, Kazaa

Six months before the first nuclear test, the Manhattan Project scientists prepared a report called LA-602. It investigated the chances of nuclear detonation having a runaway effect and destroying the Earth by burning up the atmosphere. This was probably the first time scientists performed an analysis to predict whether humanity would perish as a result of a new technological capability—the first piece of existential-risk research.

Of course, nuclear technology did not remain the last dangerous technology that humans have invented. Since then, the topic of catastrophic side effects has repeatedly come up in different contexts: recombinant DNA, synthetic viruses, nanotechnology, and so on. Luckily for humanity, sober analysis has usually prevailed and resulted in various treaties and protocols to steer the research.

When I think about the machines that can think, I think of them as technology that needs to be developed with similar (if not greater!) care. Unfortunately, the idea of AI safety has been more challenging to popularize than, say, biosafety, because people have rather poor intuitions when it comes to thinking about nonhuman minds. Also, if you think about it, AI is really a metatechnology: technology that can develop further technologies, either in conjunction with humans or perhaps even autonomously, thereby further complicating the analysis.

That said, there has been encouraging progress over the last few years, exemplified by the initiatives of new institutions, such as the Future of Life Institute, which have assembled leading AI researchers to explore appropriate research agendas, standards, and ethics.

Therefore, complicated arguments by people trying to sound clever on the issue of AI thinking, consciousness, or ethics are often a distraction from the trivial truth: The only way to ensure that we don't accidentally blow ourselves up with our own technology (or metatechnology) is to do our homework and take relevant precautions—just as those Manhattan Project scientists did when they prepared LA-602. We need to set aside the tribal quibbles and ramp up the AI safety research.

By way of analogy: Since the Manhattan Project, nuclear scientists have moved on from increasing the power extracted from nuclear fusion to the issue of how to best contain it—and we don't even call that nuclear ethics.

We call it common sense.

WHAT DO YOU CARE WHAT OTHER MACHINES THINK?

GEORGE CHURCH

Professor of genetics, Harvard Medical School; director, Harvard University's Personal Genome Project; coauthor (with Ed Regis), *Regenesis: How Synthetic Biology Will Reinvent Nature and Ourselves*

I am a machine that thinks, made of atoms—a perfect quantum simulation of a many-body problem—a 10^{29}-body problem. I, robot, am dangerously capable of self-reprogramming and preventing others from cutting off my power supply. We human machines extend our abilities via symbiosis with other machines—expanding our vision to span wavelengths beyond the mere few nanometers visible to our ancestors, out to the full electromagnetic range, from picometer to megameter. We hurl 370 kg hunks of our hive past the sun at 252,792 km/hr. We extend our memory and math by a billionfold with our silicon prostheses. Yet our biobrains are a thousandfold more energy efficient than our inorganic brains at tasks where we have common ground (like facial recognition and language translation) and infinitely better for tasks of as yet unknown difficulty, like Einstein's *Annus Mirabilis* papers or out-of-the-box inventions affecting future centuries. As Moore's Law heads from 20 nm transistor lithography down to 0.1 nm atomic precision and from 2-D to 3-D circuits, we may downplay reinventing and simulating our biomolecular brains and switch to engineering them.

We can back up petabytes of sili-brains perfectly in seconds, but transfer of information between carbo-brains takes decades, and the similarity between the copies is barely recognizable. Some speculate that we could translate from carbo to sili, and even get the sili version to behave like the original. However, such a task requires much deeper understanding than merely making a copy. We harnessed the immune system via vaccines in tenth-century China and eighteenth-century Europe, long before we understood cytokines and T-cell receptors. We don't yet have a medical nanorobot of comparable agility or utility. It may turn out that making a molecularly adequate copy of a 1.2 kg brain (or 100 kg body) is easier than understanding how it works (or than copying my brain to a room of students "multi-tasking" with smartphone cat videos and e-mails). This is far more radical than human cloning, yet doesn't involve embryos.

What civil rights issues arise with such hybrid machines? A biobrain of yesteryear with nearly perfect memory, which could reconstruct a scene with vivid prose, paintings, or animation, was permissible, often revered. But we hybrids (mutts) today, with better memory talents, are banned from courtrooms, situation rooms, bathrooms, and "private" conversations. License plates and faces are blurred in Google Street View—intentionally inflicting prosopagnosia. Should we disable or kill Harrison Bergeron? What about votes? We're currently far from universal suffrage. We discriminate based on maturity and sanity. If I copy my brain/body, does it have a right to vote or is it redundant? Consider that the copies begin to diverge immediately, or that the copy could be intentionally different. In addition to passing the maturity/sanity/humanity test, perhaps the copy needs to pass a reverse Turing Test (a Church-Turing Test?). Rather than demonstrating behavior indistinguishable from that of a human, the goal would be to show behavior distinct from human individuals. (Would the current U.S. two-party system pass such a test?) Perhaps the day of corporate personhood (*Dartmouth College v. Woodward*, 1819) has finally arrived. We already vote with our wallets. Shifts in purchasing trends result in differential wealth, lobbying, R&D priorities, etc. Perhaps more copies of specific memes, minds, and brains will come to represent the will of We the (hybrid) People of the world. Would such future Darwinian selection lead to disaster or to higher emphasis on humane empathy; aesthetics; elimination of poverty, war, and disease; long-term planning—evading existential threats on even millennial time frames? Perhaps the hybrid-brain route is not only more likely but also safer than either a leap to unprecedented, unevolved, purely silicon-based brains or sticking to our ancient cognitive biases with fear-based, fact-resistant voting.

MACHINES CANNOT THINK

ARNOLD TREHUB
Psychologist, University of Massachusetts, Amherst;
author, *The Cognitive Brain*

Machines (humanly constructed artifacts) cannot think, because no machine has a point of view—that is, a unique perspective on the worldly referents of its internal symbolic logic. We, as conscious cognitive observers, look at the output of so-called thinking machines and provide our own referents to the symbolic structures spouted by the machine. Of course, despite this limitation, such nonthinking machines have provided an extremely important adjunct to human thought.

NO "I" AND NO CAPACITY FOR MALICE

ROY BAUMEISTER

Francis Eppes Eminent Scholar and head, social psychology graduate program, Florida State University; coauthor (with John Tierney), *Willpower: Rediscovering the Greatest Human Strength*

So-called thinking machines are extensions of the human mind. They don't exist in nature. They're not created by evolution, they're created by human minds from blueprints and theories. The human mind figures out how to make tools that enable it to work better. A computer is one of the best tools.

Life mostly seeks to sustain life, so living things care about what happens. The computer, not alive and not designed by evolution, doesn't care about survival or reproduction—in fact, it doesn't care about anything. Computers aren't dangerous in the way snakes and hired killers are dangerous. Although many movies explore horror fantasies of computers turning malicious, real computers lack the capacity for malice.

A thinking machine that serves a human is an asset, not a threat. Only if it became an independent agent, acting on its own—a tool rebelling against its user's wishes—could it become a threat. For that, a computer would need to do more than think. It would need to make choices that could violate the programmer's wishes. That would require something akin to free will.

What would the computer on your desk or lap have to do so that you'd say it had free will (at least in whatever sense humans have free will)? Certainly it would have to be able to reprogram itself; otherwise it would just be carrying out built-in instructions. And the reprogramming would have to be done in a way that was flexible, not programmed in advance. But where would that come from? In humans, the agent comes to exist because it serves the motivational system: It helps you get what you need and want.

Humans, like other animals, were designed by evolution, and so the beginnings of subjectivity come with wanting and liking the things that enable life to continue, like food and sex. The agent serves that, choosing actions that obtain those life-sustaining things. And thinking helps the agent make better choices.

Human thinking thus serves to prolong life—by helping you decide who to trust, what to eat, how to make a living, who to marry. A thinking machine isn't motivated by any innate drive to sustain its life. The computer may be able to process more information faster than a human brain can, but there's no "I" in the computer, because it doesn't begin with wanting things that enable it to sustain life. If computers had an urge to prolong their own existence, they'd probably focus their ire mainly on the computer industry so as to stop progress. Because the main threat to a computer's continued existence arises when newer, better computers make it obsolete.

LEVERAGING HUMAN INTELLIGENCE

KEITH DEVLIN
Mathematician; executive director, H-STAR Institute, Stanford University; author, *The Man of Numbers: Fibonacci's Arithmetic Revolution*

I know many machines that think. They're people. Biological machines. (Be careful of that last phrase, "biological machines." It's a convenient way to refer to stuff we don't fully understand in a way that suggests we do.) In contrast, I've yet to encounter a digital-electronic, electromechanical machine that behaves in a fashion that would merit the description "thinking," and I see no evidence to suggest that such may even be possible. HAL-like devices that will eventually rule us are, I believe, destined to remain in the realm of science fiction. Just because something waddles like a duck and quacks doesn't make it a duck. And if a machine exhibits some features of thinking (e.g., decision making), that doesn't make it a thinking machine.

We humans are suckers for being seduced by the "if it waddles and quacks, it's a duck" syndrome. Not because we're stupid; rather, because we're human. The very features that allow us to act most of the time in our best interests when faced with potential information overload in complex situations leave us wide open for such seduction.

Many years ago, I visited a humanoid robotics lab in Japan. It looked like a typical engineering Skunk Works. In one corner was a metallic skeletal device, festooned with electrical wires, which had the rough outline of a human upper torso.

The sophisticated-looking arms and hands were, I assume, the focus of much of the engineering research, but they weren't active during my visit, and it was only later that I really noticed them. My entire attention when I walked in, and for much of my time there, was taken up by the robot's head. Actually, it wasn't a head at all, just a metal frame with a camera where the nose and mouth would be. Above the camera were two white balls, about the size of Ping-Pong balls (which may be what they were), with black pupils painted on them. Above the eyeballs were two large paperclips, serving as eyebrows.

The robot was programmed to detect the motion of people and pick up sound sources (from someone speaking). It would move its head and its eyeballs to follow anyone who moved, and it would raise and lower its paperclip eyebrows when the target individual was speaking.

What was striking was how alive and intelligent the device seemed. Sure, I and everyone else in the room knew exactly what was going on and how simple the mechanism was that controlled the robotic "gaze" and the paperclip eyebrows. It was a trick. But it was a trick that tapped deep into hundreds of thousands of years of human social and cognitive development, so our natural response was the one normally elicited by another person.

It wasn't even that I was unaware of how the trick worked. My then Stanford colleague and friend, the late Clifford Nass, had done hundreds of hours of research showing how we humans are genetically programmed to ascribe intelligent agency based on a few simple interaction clues—reactions so deep and ingrained that we cannot eliminate them. There probably was some sophisticated AI that could control the robot's arms and hands, but the eyes and eyebrows were controlled by a very simple program. Even so, that behavior was enough to give me the clear sense that the robot was a curious, intelligent participant, able to follow what I said. What it was doing, of course, was leveraging *my* humanity and *my* intelligence. It wasn't thinking.

Leveraging human intelligence is all well and good if the robot is to clean your house, book your airline tickets, or drive your car. But would you want such a machine to serve on a jury, make a crucial decision regarding a hospital procedure, or have control over your freedom? I certainly wouldn't.

So when you ask me what I think about machines that think, my answer is that for the most part I like them, because they're people (and perhaps also various other animals). What worries me is the increasing degree to which we're ceding aspects of our lives to machines that *decide*, often much more effectively and reliably than people can, but definitely don't *think*. There's the danger: machines that can make decisions but don't think.

Decision making and thinking aren't the same, and we shouldn't confuse them. When we deploy decision-making systems in matters of national defense, health care, and finance, as we do, the potential dangers of such confusion, both for individuals and for society, are particularly high. To guard against those dangers, it helps to be aware that we're genetically programmed to act in trustful, intelligent-agency-ascribing ways in certain kinds of interactions, be they with people or machines. But sometimes a device that waddles and quacks is just a device. It ain't no duck.

A MACHINE IS A "MATTER" THING

EMANUEL DERMAN
Professor of financial engineering, Columbia University; senior adviser, KKR Prisma; author, *Models.Behaving.Badly* and *My Life As a Quant*

A machine is a small part of the physical universe that has been arranged, after some thought by humans or animals, in such a way that when certain initial conditions are set up (by humans or animals) the deterministic laws of nature see to it that that small part of the physical universe automatically evolves in a way that humans or animals think is useful.

A machine is a "matter" thing that gets its quality from the point of view of a "mind." There is a "mind" way of looking at things and a "matter" way of looking at things.

Stuart Hampshire, in *Spinoza: An Introduction to His Philosophical Thought* (1988), argued that according to Spinoza, you must choose: You can invoke mind as an explanation for something mind-like, or you can invoke matter as an explanation for something material, but you cannot fairly invoke mind to explain matter, or vice versa. In Hampshire's example, suppose you become embarrassed and turn red. You might commonly say, "I blushed because I became embarrassed." A strict Spinozist, according to Hampshire, would not claim that embarrassment was the cause of blushing, because embarrassment is the mental description and the blush is physical, and you should not crisscross your causal chains. That would be sloppy thinking. Embarrassment and blushing are complementary, not causal.

By this argument, one should not jump from one style of explanation to another. We must explain physical things by physics and psychological things by psychology. It is of course very difficult to give up the notion of psychic causes of physical states or physical causes of psychic states.

So far, I like this view of the world. I will therefore describe mental behavior in mental terms (lovesickness made me moody) and material behavior by material causes (drugs messed up my body chemistry). From this point of view therefore, as long as I understand the material explanation of a machine's behavior, I will argue that it doesn't think.

I realize I may have to change this view when someone genuinely does away with the complementary view of mind and matter and convincingly puts matter as the cause of mind or mind as the cause of matter. So far, though, this is just an issue of faith.

Until then—and maybe that day will come, but as yet I see no sign of it—I think that machines can't think.

I COULD BE WRONG

FREEMAN DYSON
Physicist, Institute for Advanced Study;
author, *Dreams of Earth and Sky*

I do not believe that machines that think exist, or that they are likely to exist in the foreseeable future. If I am wrong, as I often am, any thoughts I might have about the question are irrelevant.

If I am right, then the whole question is irrelevant.

WHY CAN'T "BEING" OR "HAPPINESS" BE COMPUTED?

DAVID GELERNTER
Computer scientist, Yale University; chief scientist, Mirror Worlds Technologies; author, *America-Lite: How Imperial Academia Dismantled our Culture (and Ushered in the Obamacrats)*

The now old-fashioned idea of "machines that think" shows a deep but natural misunderstanding of the mind and software. Computers will never think; to see why, let's start with french fries. I'm assuming that the machines in question are computers, but a variant of the argument applies to any machine.

Neither *french fries* nor *french fried* is computable—no computer can ever produce *french fries* as a result, or the *french fried* state of being. *French fried* is not computable because it's a physical state of a particular object, and computers produce only information or codes for information, not physical stuff, not transformations of physical stuff. *Happy* is also a physical state of a particular object—namely, a person. *Happy* can't exist unless you start with a person and put him into a state of happiness. Computers can't do that.

Thinking-about and *being*, or (equivalently) *thinking-about* and *feeling*, are the endpoints of a spectrum that defines the human mind. (By *feeling*, I mean *sensation*, *emotion*, or *mood*, just as the English language does.) We need the whole spectrum or we have no mind and no thought in any proper sense. Computers can imitate important aspects of *thinking-about* (narrowly understood), but *being* is beyond them. Therefore mindfulness is beyond them.

The word *being* is a useful abbreviation, in this context, for *being part of a physical object or system, and responding naturally to that environment*. A sliced potato can be part of a fryer environment and respond by turning french-fried. Litmus paper can be part of an acid-in-a-beaker system and respond by turning blue. The mind is like litmus paper, but instead of turning colors, it responds to its surroundings by experiencing them. If something gives us grounds to be happy, the mind-body system (the human being) becomes happy, and the mind experiences happiness. Happiness has mental and physical consequences. You might experience a rush of energy, even quickened pulse and breathing.

Why can't *being* be computed? Why can't *happiness*? Happiness is not computable because, as the state of a physical object, it's outside the universe of computation. Computers and software don't create or manipulate physical stuff. They can cause other, attached machines to do that, but what those attached machines do isn't the accomplishment of computers. Robots can fly, but computers can't. (Nor is any computer-controlled device guaranteed to make people happy—but that's another story.) *Being* is not computable—an important fact that's been overlooked until now, not surprisingly. Computers and the mind live in different universes, like pumpkins and Puccini, and are hard to compare whatever one intends to show.

Can we get by without *being* and still have a thinking machine? No. *Thinking-about* and *being* (or *feeling*) define the mind and its capacities. At the spectrum's top—at maximum alertness or focus—the mind throws itself into *thinking-about* and fends off emotion, which is distracting. At the spectrum's bottom is sleep-and-dreaming, a state in which we do little thinking; we're preoccupied by sensation as we hallucinate, and often by emotion (dreams can be strongly emotional)—in any event, with *feeling*, or (in other words) *being*.

Why did it take so long to produce such a simple argument? And why are so few thinkers likely to accept it now? Maybe because most philosophers and scientists wish that the mind *were* nothing but thinking and that *feeling* or *being* played no part. They wished so hard for it to be true that they finally decided it was. Philosophers are only human.

NO MACHINE THINKS ABOUT THE ETERNAL QUESTIONS

LEO M. CHALUPA
Neurobiologist; vice president for research,
George Washington University

Recent demonstrations of the prowess of high-performance computers are remarkable but unsurprising. With proper programming, machines are far superior to humans in storing and assessing vast quantities of data and making virtually instantaneous decisions. These are machines that think, because similar processes are involved in much of human thought.

But in a broader sense, the term *thinking machine* is a misnomer. No machine has ever thought about the eternal questions: Where did I come from? Why am I here? Where am I going? Machines don't think about their future, their ultimate demise, or their legacy. To ponder such questions requires consciousness and a sense of self. Thinking machines don't have these attributes, and given the current state of our knowledge they're unlikely to attain them in the foreseeable future.

The only viable approach to constructing a machine that has the attributes of the human brain is to copy the neuronal circuits underlying thinking. Indeed, research programs now ongoing at UC Berkeley, MIT, and several other universities are focused on achieving this objective, striving to build computers that function like the cerebral cortex. Recent advances in our understanding of cortical microcircuitry have propelled this work, and it's likely that the recent White House BRAIN Initiative will provide a wealth of valuable additional information. In the coming decades, we'll know how the billions of neurons in each of the six layers of the cerebral cortex are interconnected, as well as the types of functional circuits these connections form.

This is a much-needed first step in designing machines capable of thinking in a manner equivalent to the human brain. But understanding the cortical microcircuitry isn't sufficient to construct a machine that thinks. What's required is an understanding of the neuronal activity underlying the thinking process. Imaging studies have revealed much new information about the brain regions involved in such functions as vision, hearing, touch, fear, pleasure, and others.

But as yet we don't have even a preliminary understanding of what takes place when we are in thought. There are many reasons for this, not the least of which is our inability to isolate the thinking process from other bodily states. Moreover, different brain circuits may well be engaged in different modes of thinking. Thinking about an upcoming lecture would be expected to activate the brain differently from thinking about unpaid bills.

In the near term, we can expect computers to do more and more things better than humans. But we need a far greater understanding of the workings of the human brain to create a machine that thinks in a way equivalent to human thought. For now, we don't need to be concerned with civil or any other rights of machines that think; nor do we need to be concerned at the prospect of thinking machines taking over society. If things should get out of hand, just pull the plug.

THE SINGULARITY—
AN URBAN LEGEND?

DANIEL C. DENNETT

Philosopher; Austin B. Fletcher Professor of Philosophy and codirector, Center for Cognitive Studies, Tufts University; author, *Intuition Pumps and Other Tools for Thinking*

The Singularity—the fateful moment when AI surpasses its creators in intelligence and takes over the world—is a meme worth pondering. It has the earmarks of an urban legend: a certain scientific plausibility ("Well, in principle I guess it's possible!") coupled with a deliciously shudder-inducing punch line ("We'd be ruled by robots!"). Did you know that if you sneeze, belch, and fart all at the same time, *you die*? Wow! Following in the wake of decades of AI hype, you might think the Singularity would be regarded as a parody, a joke, but it has proved to be a remarkably persuasive escalation. Add a few illustrious converts—Elon Musk, Stephen Hawking, and David Chalmers, among others—and how can we not take it seriously? Whether this stupendous event occurs 10 or 100 or 1,000 years in the future, isn't it prudent to start planning now, setting up the necessary barricades and keeping our eyes peeled for harbingers of catastrophe?

I think, on the contrary, that these alarm calls distract us from a more pressing problem, an impending disaster that won't need any help from Moore's Law or further breakthroughs in theory to reach its much closer tipping point: After centuries of hard-won understanding of nature that now permits us, for the first time in history, to control many aspects of our destinies, we're on the verge of abdicating this control to artificial agents that *can't* think, prematurely putting civilization on autopilot. The process is insidious, because each step of it makes good local sense, is an offer you can't refuse. You'd be a fool today to do large arithmetical calculations with pencil and paper when a hand calculator is much faster and almost perfectly reliable (don't forget about round-off error), and why memorize train timetables when they're instantly available on your smartphone? Leave the map reading and navigation to your GPS; it isn't conscious, it can't think in any meaningful sense, but it's much better than you are at keeping track of where you are and where you want to go.

Much farther up the staircase, doctors are becoming increasingly dependent on diagnostic systems that are provably more reliable than any human diagnostician. Do you want *your* doctor to overrule the machine's verdict when it comes to making a lifesaving choice of treatment? This may prove to be the best—most provably successful, most immediately useful—application of the technology behind IBM's Watson, and the issue of whether or not Watson can properly be said to think (or be conscious) is beside the point. If Watson turns out to be better than human experts at generating diagnoses from available data, we'll be morally obliged to avail ourselves of its results. A doctor who defies it will be asking for a malpractice suit. No area of human endeavor appears to be clearly off-limits to such prosthetic performance-enhancers, and wherever they prove themselves, the forced choice will be reliable results over the human touch, as it always has been. Handmade law and even science could come to occupy niches adjacent to artisanal pottery and hand-knit sweaters.

In the earliest days of AI, an attempt was made to enforce a sharp distinction between *artificial intelligence* and *cognitive simulation*. The former was to be a branch of engineering, getting the job done by hook or by crook, with no attempt to mimic human thought processes—except when that proved to be an effective way of proceeding. Cognitive simulation, in contrast, was to be psychology and neuroscience conducted by computer modeling. A cognitive-simulation model that nicely exhibited recognizably human errors or confusions would be a triumph, not a failure. The distinction in aspiration lives on, but has largely been erased from public consciousness: To laypeople, AI means passing the Turing Test, being humanoid. The recent breakthroughs in AI have been largely the result of turning away from (what we thought we understood about) human thought processes and using the awesome data-mining powers of supercomputers to grind out valuable connections and patterns *without* trying to make them understand what they're doing. Ironically, the impressive results are inspiring many in cognitive science to reconsider; it turns out that there's much to learn about how the brain does its brilliant job of "producing future" by applying the techniques of data mining and machine learning.

But the public will persist in imagining that any black box that can do *that* (whatever the latest AI accomplishment is) must be an intelligent agent much like a human being, when in fact what's inside the box is a bizarrely truncated, two-dimensional fabric that gains its power precisely by not adding the overhead of a human mind, with all its distractability, worries, emotional commitments, memories, allegiances. It's not a humanoid robot at all but a mindless slave, the latest advance in autopilots.

What's wrong with turning over the drudgery of thought to such high-tech marvels? Nothing, so long as (1) we don't delude ourselves, and (2) we somehow manage to keep our own cognitive skills from atrophying.

1. It is very, very hard to imagine (and keep in mind) the limitations of entities that can be such valued assistants, and the human tendency is always to overendow them with understanding—as we have known since Joseph Weizenbaum's notorious Eliza program of the 1960s. This is a huge risk, since we'll always be tempted to ask more of them than they were designed to accomplish, and to trust the results when we shouldn't.
2. Use it or lose it. As we become ever more dependent on these cognitive prostheses, we risk becoming helpless if they ever shut down. The Internet is not an intelligent agent (well, in some ways it is), but we have nevertheless become so dependent on it that were it to crash, panic would set in and we could destroy society in a few days. That's an event we should bend our efforts to averting *now,* because it could happen any day.

The real danger, then, is not machines that are more intelligent than we are usurping our role as captains of our destinies. The real danger is basically clueless machines being ceded authority far beyond their competence.

NANO-INTENTIONALITY

W. TECUMSEH FITCH
Professor of cognitive biology, University of Vienna;
author, *The Evolution of Language*

Despite vast increases in computing power, current computers don't think the way we do (or a chimpanzee or a dog does). Silicon-based computers lack a crucial capacity of organic minds: the ability to change their detailed material form, and thus their future computations, in response to events in the world. Without this ability (which elsewhere I've dubbed *nano-intentionality*), information processing alone doesn't amount to meaningful thought, because the symbols and values being computed lack any intrinsic causal connection to the real world. Silicon-based information processing requires interpretation by humans to become meaningful and will for the foreseeable future. We have little to fear from thinking machines and more to fear from the increasingly unthinking humans who use them.

What exactly is this property present in biological but not silicon computers? Fear not that I'm invoking some mystical élan vital: This is an observable, mechanistic property of living cells—a property that evolved via normal Darwinian processes. No mysticism or "invisible spirit" lurks in my argument. At its heart, nano-intentionality is the ability of cells to respond to changes in their environment by rearranging their molecules and thus changing their form. It's present in an amoeba engulfing a bacterium, a muscle cell boosting myosin levels in response to jogging, or (most relevant) a neuron extending its dendrites in response to its local neurocomputational environment. Nano-intentionality is a basic, irreducible, undeniable feature of life on Earth, and is *not* present in the engraved, rigid silicon chips forming the hearts of modern computers. Because this physical difference between brains and computers is a simple brute fact, the issue open to debate is what significance this fact has for more abstract philosophical issues concerning "thought" and "meaning." This is where the argument gets more complicated.

The philosophical debate starts with Kant's observation that our minds are irrevocably separated from the typical objects of our thoughts—physical entities in the world. We gather evidence about these objects (via photons or air vibrations or molecules they release) but our minds/brains never make direct contact with them. Thus, the question of how our mental entities (thoughts, beliefs, desires) can be said to be "about" things in the real world is surprisingly problematic. Indeed, this problem of "aboutness" is a central problem in the philosophy of mind, at the heart of decades-long debates between philosophers like Dennett, Fodor, and Searle. Philosophers have rather unhelpfully dubbed this putative mental "aboutness" *intentionality* (not to be confused with the everyday English meaning, "doing something on purpose"). Issues of intentionality are closely tied with deep issues about phenomenal consciousness, often framed in terms of "qualia" and the "hard problem" of consciousness, but they address a more basic and fundamental question: How can a mental entity (a thought—a pattern of neural firing) be in any sense "connected" to its object (a thing you see or the person you're thinking about)?

The skeptical, solipsistic answer is: There is no such connection; intentionality is an illusion. This conclusion is false in at least one crucial domain (highlighted by Schopenhauer 200 years ago): The one place where mental events (desires and intentions, as instantiated in neural firing) make contact with the "real world" is *within our own bodies* (e.g., at the neuromuscular junction). In general, the plasticity of living matter, and neurons in particular, means that a feedback loop directly connects our thoughts to our actions, percolating back through our perceptions to influence the structure of neurons themselves. This loop is closed every day in our brains. (Indeed, if you remember anything about this essay tomorrow, it's because some neurons in your brain changed their form, weakening or strengthening synapses, extending or withdrawing connections.) This feedback loop cannot in principle be closed in a rigid silicon chip. This biological quality grants our mental activities (or a chimpanzee's or dog's) with a causal intrinsic intentionality lacking in contemporary silicon computing systems.

To the extent that this argument is correct (and both logic and intuition support it), machines "think," "know," or "understand" only insofar as their makers and programmers do, when meaning is added by an intentional, interpreting agent with a brain. Any "intelligence" of AIs is derived solely from their creators.

I thus have no fear of an AI uprising or AI rights movement (except perhaps for one led by deluded humans). Does this mean we're in the clear until someone eventually designs a computer with nano-intentionality? Unfortunately not; there's a different danger created by our strong anthropomorphic tendency to misattribute intentions and understanding to inanimate objects ("my car dislikes low-octane fuel"). When we apply this to computational artifacts (computers, smartphones, control systems), there's a strong tendency to gradually cede our responsibilities—informed, competent understanding—to computers (and those who control them). Danger begins when we willingly and lazily cede this unique competence to myriad silicon systems (car navigators, smartphones, electronic voting systems, the global financial system) that neither know nor care what they're computing about. The global financial crisis gave us a taste of what's possible in a computer-interconnected world when responsibility and competence have unwisely been offloaded to machines.

I don't fear the triumphal uprising of AIs but, rather, a catastrophic system failure caused by multiple minor bugs in overempowered, interconnected silicon systems. We remain far from any Singularity in which computers outsmart us, but this is no insurance against a disastrous network collapse. The first step in avoiding such catastrophes is to stop granting computers responsibility for meaningful thought or understanding and accept a basic truth: Machines don't think. And thinking they do becomes riskier every day.

A BEAUTIFUL (VISIONARY) MIND

IRENE PEPPERBERG
Research associate and lecturer, Psychology Department, Harvard University; author, *Alex & Me*

While machines are terrific at computing, they're not very good at actual thinking. Machines have an endless supply of grit and perseverance, and, as others have said, will effortlessly crunch out the answer to a complicated mathematical problem or direct you through traffic in an unknown city, all by use of the algorithms and programs installed by humans. But what do machines lack?

Machines (at least so far, and I don't think this will change with a Singularity) lack *vision*. And I don't mean sight. Machines do not devise the next new killer app on their own. Machines don't decide to explore distant galaxies—they do a terrific job once we send them, but that's a different story. Machines are certainly better than the average person at solving problems in calculus and quantum mechanics—but machines don't have the vision to see the need for such constructs in the first place. Machines can beat humans at chess, but they have yet to design the type of mind game that will intrigue humans for centuries. Machines can see statistical regularities my feeble brain will miss, but they can't make the insightful leap that connects disparate sets of data to devise a new field.

I'm not terribly concerned about machines that compute. I'll deal with the frustration of my browser in exchange for a smart refrigerator that, based on tracking RFID codes of what comes in and out, texts me to buy cream on my way home (hint to those working on such a system . . . sooner rather than later!). I like having my computer underline words it doesn't recognize, and I'll deal with the frustration of having to ignore its comments on "phylogenetic" in exchange for catching my typo on a common term (in fact, it won't let me misspell a word here to make a point). But these examples show that just because a machine goes through the motions of what looks like thinking doesn't mean it's actually engaging in that behavior—or at least, in one equivalent to the human process.

I'm reminded of one of the earliest studies to train apes to use "language"—in this case, to manipulate plastic chips to answer a number of questions. The system was replicated with college students, who—not surprisingly—did exceptionally well but, when asked what they'd been trained to do, claimed that they'd solved some interesting puzzles and had no idea they were being taught a language. Much debate ensued, and much was learned and put into practice in subsequent studies, so that several nonhuman subjects did eventually understand the referential meaning of the various symbols they were taught to use, and we did learn a lot about ape intelligence from the original methodology. The point, however, is that what initially looked like a complicated linguistic system needed a lot more work before it became more than a series of (relatively) simple paired associations.

My concern therefore is not about thinking machines but about a complacent society—one that might give up on its visionaries in exchange merely for getting rid of drudgery. Humans need to take advantage of the cognitive capacity that's freed when machines take over the scut work—and be thankful for that freedom and *use* that freedom, channeling all that ability into the hard work of solving pressing problems that need insightful, visionary leaps.

THE COLOSSUS IS A BFG

NICHOLAS HUMPHREY

Emeritus professor of psychology, London School of Economics; visiting professor of philosophy, New College of the Humanities; senior member, Darwin College, University of Cambridge; author, *Soul Dust: The Magic of Consciousness*

A penny for your thoughts? You may not choose to answer, but the point is that as a conscious agent, you can. That's what it means to have introspective access. You know—and can tell us—what's on stage in the theater of your mind. Then how about machines? A bitcoin for the thinking machine's thoughts? But no one has yet designed a machine to have that kind of access. Wittgenstein remarked that if a lion could speak, we wouldn't understand him. If a machine could speak, it wouldn't have anything to say. What do I think about machines that think? Simple. I don't think there are, as yet, any such machines.

Of course, this may soon change. Far back in human history, natural selection discovered that given the particular problems humans faced, there were practical advantages to having a brain capable of introspection. Likewise, machine programmers may well discover that when and if machines face similar problems, the software trick that works for humans will work for them too. But what are these problems, and why is the theater of consciousness the answer?

The theater lets you in on a secret: It lets you see how your own mind works. Observing, for example, how beliefs and desires generate wishes that lead to actions, you begin to gain insight into why you think and act the way you do. So you can explain yourself to yourself and to other people too. But equally important, it means you have a model for explaining other people to yourself. Introspective consciousness has laid the ground for what psychologists call Theory of Mind.

With humans, for whom social intelligence is the key to biological survival, the advantages have been huge. With machines, for whom success in social life has not yet become an issue, there has been little if any reason to go that way. However, there's no question that the time is coming when machines will indeed need to understand other machines' psychology, so as to be able to work alongside them. What's more, if they're to collaborate effectively with humans, they'll need to understand human psychology too. I guess that's when their designers—or maybe the machines themselves—will follow nature's lead and install a machine version of the inner eye.

Is there a danger that once this stage is reached, these newly insightful machines will come to understand humans only too well? Psychopaths are sometimes credited with having not too little but too great an understanding of human psychology. Is this something we should fear with machines?

I don't think so. This situation is not a new one. For thousands of years, humans have been selecting and programming a particular species of biological machine to act as servants, companions, and helpmeets to ourselves. I'm talking of the domestic dog. The remarkable result has been that modern dogs have in fact acquired an exceptional and considerable ability to mind-read—both the minds of other dogs and of humans—superior to that of any animal other than humans themselves. This has evidently evolved as a mutually beneficial relationship, not a competition, even if it's one in which we have retained the upper hand. If and when it gets to the point where machines are as good at reading human minds as dogs now are, we shall of course have to watch out in case they get too dominant and manipulative, perhaps even too playful—just as we already have to do with man's best friend. But I see no reason to doubt we'll remain in control.

There's a painting by Goya of a terrible Colossus who strides across the landscape while the human population flees in terror. Colossus was the name of one of Alan Turing's first computing machines. Do we have to imagine an existential threat to humanity coming from that computer's descendants? No, I look on the bright side. With luck, or rather by arrangement, the Colossus will remain a Big Friendly Giant.

SELF-AWARE AI?
NOT IN 1,000 YEARS!

ROLF DOBELLI
Founder, Zurich Minds; journalist;
author, *The Art of Thinking Clearly*

The widespread fear that AI will endanger humanity and take over the world is irrational. Here's why.

Conceptually, autonomous or artificial intelligence systems can develop in two ways: either as an extension of human thinking or as radically new thinking. Call the first "Humanoid Thinking," or Humanoid AI, and the second "Alien Thinking," or Alien AI.

Almost all AI today is Humanoid Thinking. We use AI to solve problems too difficult, time-consuming, or boring for our limited brains to process: electrical-grid balancing, recommendation engines, self-driving cars, face recognition, trading algorithms, and the like. These artificial agents work in narrow domains with clear goals their human creators specify. Such AI aims to accomplish human objectives—often better, with fewer cognitive errors, distractions, outbursts of bad temper, or processing limitations. In a couple of decades, AI agents might serve as virtual insurance sellers, doctors, psychotherapists, and maybe even virtual spouses and children.

But such AI agents will be our slaves, with no self-concept of their own. They'll happily perform the functions we set them up to do. If screwups happen, they'll be *our* screwups, due to software bugs or overreliance on these agents (Dan Dennett's point). Yes, Humanoid AIs might surprise us once in a while with novel solutions to specific optimization problems. But in most cases novel solutions are the last thing we want from AI (creativity in nuclear-missile navigation, anyone?). That said, Humanoid AI solutions will always fit a narrow domain. They'll be understandable, either because we understand what they achieve or because we understand their inner workings. Sometimes the code will become too enormous and fumbled for one person to understand, because it's continually patched. In these cases, we can turn it off and program a more elegant version. Humanoid AI will bring us closer to the age-old aspiration of having robots do most of the work while humans are free to be creative—or amused to death.

Alien Thinking is radically different. Alien Thinking could conceivably become a danger to Humanoid Thinking; it could take over the planet, outsmart us, outrun us, enslave us—and we might not even recognize the onslaught. What sort of thinking will Alien Thinking be? By definition, we can't tell. It will encompass functionality we cannot remotely understand. Will it be conscious? Most likely, but it needn't be. Will it experience emotion? Will it write bestselling novels? If so, bestselling to us or bestselling to it and its spawn? Will cognitive errors mar its thinking? Will it be social? Will it have a Theory of Mind? If so, will it make jokes, will it gossip, will it worry about its reputation, will it rally around a flag? Will it create its own version of AI (AI-AI)? We can't say.

All we can say is that humans cannot construct truly Alien Thinking. Whatever we create will reflect our goals and values, so it won't stray far from human thinking. You'd need real evolution, not just evolutionary algorithms, for self-aware Alien Thinking to arise. You'd need an evolutionary path radically different from the one that led to human intelligence and Humanoid AI.

So, how do you get real evolution to kick in? Replicators, variation, and selection. Once these three components are in place, evolution arises inevitably. How likely is it that Alien Thinking will evolve? Here's a back-of-the-envelope calculation:

First, consider what getting from magnificently complex eukaryotic cells to human-level thinking involved. Achieving human thought required a large part of the Earth's biomass (roughly 500 billion tons of eukaryotically bound carbon) during approximately 2 billion years. That's a lot of evolutionary work! True, human-level thinking might have happened in half the time. With a lot of luck, even in 10 percent of the time, but it's unlikely to have happened any faster. You don't only need massive amounts of time for evolution to generate complex behavior, you also need a petri dish the size of Earth's surface to sustain this level of experimentation.

Assume that Alien Thinking will be silicon-based, as all current AI is. A eukaryotic cell is vastly more complex than, say, Intel's latest i7 CPU chip—both in hardware and software. Further assume that you could shrink that CPU chip to the size of a eukaryote. Leave aside the quantum effects that would stop the transistors from working reliably. Leave aside the question of the energy source. You'd have to cover the globe with 10^{30} microscopic CPUs and let them communicate and fight for 2 billion years for true thought to emerge.

Yes, processing speed is faster in CPUs than in biological cells, because electrons are easier to shuttle around than atoms. But eukaryotes work massively parallel, whereas Intel's i7 works only four times parallel (four cores). Eventually, at least to dominate the world, these electrons would need to move atoms to store their software and data in more and more physical places. This would slow their evolution dramatically. It's hard to say if, overall, silicon evolution will be faster than biological. We don't know enough about it. I don't see why this sort of evolution would be more than two or three orders of magnitude faster than biological evolution (if at all)—which would bring the emergence of self-aware Alien AI down to roughly a million years.

What if Humanoid AI becomes so smart it could create Alien AI from the top down? That's where Leslie Orgel's Second Rule kicks in: "Evolution is smarter than you are." It's smarter than human thinking. It's even smarter than Humanoid Thinking. And it's much slower than you think.

Thus, the danger of AI is not inherent to AI but rests on our overreliance on it. Artificial thinking won't evolve to self-awareness in our lifetime. In fact, it won't happen in 1,000 years.

I might be wrong, of course. After all, this back-of-the-envelope calculation applies legacy human thinking to Alien AI—which by definition we won't understand. But that's all we can do at this stage.

Toward the end of the 1930s, Samuel Beckett wrote in a diary, "We feel with terrible resignation that reason is not a superhuman gift . . . that reason evolved into what it is, but that it also, however, could have evolved differently." Replace "reason" with "AI" and you have my argument.

MACHINES DON'T THINK, BUT NEITHER DO PEOPLE

CESAR HIDALGO

Associate professor, MIT Media Lab; author, *Why Information Grows: The Evolution of Order, from Atoms to Economies*

Machines that think? That's as fallacious as people who think! Thinking involves processing information, begetting new physical order from incoming streams of physical order. Thinking is a precious ability, which unfortunately is not the privilege of single units such as machines or people but a property of the systems in which these units come to "life."

Of course I'm being provocative here, since at the individual level we do process information. We do think—sometimes—or at least we feel like we do. But "our" ability to think is not entirely "ours"—it's borrowed, since the hardware and software we use to think weren't begot by us. You and I did not evolve the genes that helped organize our brains or the language we use to structure our thoughts. Our ability to think is dependent on events that happened prior to our mundane existence: the past chapters of biological and cultural evolution. So we can only understand our ability to think, and the ability of machines to mimic thought, by considering how the ability of a unit to process information relates to its context.

Think of a human born in the dark solitude of empty space. She'd have nothing to think about. The same would be true of an isolated and inputless computing machine. In this context, we can call our borrowed ability to process information "little" thinking—since it's a context-dependent ability that happens at the individual level. "Large" thinking, by contrast, is the ability to process information embodied in systems, where units like machines or us are mere pawns.

Separating the little thinking of humans from the larger thinking of systems (which involves the process that begets the hardware and software that allow units to "little think") helps us understand the role of thinking machines in this larger context. Our ability to think isn't only borrowed; it also hinges on the use and abuse of mediated interactions. For human/machine systems to think, humans need to eat and regurgitate one another's mental vomit, which sometimes takes the form of words. But since words vanish in the wind, our species' enormous ability to think hinges on more sophisticated techniques to communicate and preserve the information we generate: our ability to encode information in matter.

For 100,000 years, our species has been busy transforming our planet into a giant tape player. The planet Earth is the medium wherein we print our ideas: sometimes in symbolic form, such as text and paintings, but, more important, in objects—like hair dryers, vacuum cleaners, buildings, and cars—built from the mineral loins of planet Earth. Our society has a great collective ability to process information because our communication involves more than words: It involves the creation of objects, which transmit not something as flimsy as an idea but something as concrete as know-how and the uses of knowledge. Objects augment us; they allow us to do things without knowing how. We all get to enjoy the teeth-preserving powers of toothpaste without knowing how to synthesize sodium fluoride, or the benefits of long-distance travel without knowing how to build a plane. By the same token, we all enjoy the benefits of sending texts throughout the world in seconds through social media or of performing complex mathematical operations by pressing a few keys on a laptop computer.

But our ability to create the trinkets augmenting us has also evolved, of course, as a result of our collective willingness to eat one another's mental vomit. This evolution is the one that brings us now to the point where we have "media" that are beginning to rival our ability to process information, or "little think."

For most of our history, our trinkets were static objects. Even our tools were solidified chunks of order, such as stone axes, knives, and knitting needles. A few centuries ago, we developed the ability to outsource muscle and motion to machines, causing one of the greatest economic expansions in history. Now we've evolved our collective ability to process information by creating objects endowed with the ability to beget and recombine physical order. These are machines that can process information—engines that produce numbers, like the engines Charles Babbage dreamed about.

So we've evolved our ability to think collectively by first gaining dominion over matter, then over energy, and now over physical order, or information. Yet this shouldn't fool us into believing that we think or that machines do. The large evolution of human thought requires mediated interactions, and the future of thinking machines will also happen at the interface where humans connect with humans through objects.

As we speak, nerds in the best universities of the world are mapping out the brain, building robotic limbs, and developing primitive versions of technologies that will open up the future when your great-grandchild will get high by plugging his brain directly into the Web. The augmentation these kids will get is unimaginable to us—and so bizarre by our modern ethical standards that we're not even in a position to properly judge it; it would be like a sixteenth-century Puritan judging present-day San Francisco. Yet in the grand scheme of the universe, these new human/machine networks will be nothing other than the next natural step in the evolution of our species' ability to beget information. Together, humans and our extensions—machines—will continue to evolve networks that are enslaved to the universe's main glorious purpose: the creation of pockets where information does not dwindle but grows.

TANGLED UP IN THE QUESTION

JAMES J. O'DONNELL
Classical scholar; University Professor, Georgetown University; author, *Augustine, The Ruin of the Roman Empire, Pagans*

Thinking is a word we apply with no discipline whatsoever to a huge variety of reported behaviors. "I think I'll go to the store," and "I think it's raining," and "I think, therefore I am," and "I think the Yankees will win the World Series," and "I think I'm Napoleon," and "I think he said he would be here, but I'm not sure"—all use the same word to mean entirely different things. Which of them might a machine do someday? I think that's an important question.

Could a machine get confused? Experience cognitive dissonance? Dream? Wonder? Forget the name of that guy over there and at the same time know that it really knows the answer and if it just thinks about something else for a while, it might remember? Lose track of time? Decide to get a puppy? Have low self-esteem? Have suicidal thoughts? Get bored? Worry? Pray? I think not.

Can artificial mechanisms be constructed to play the part in gathering information and making decisions that human beings now play? Sure, they already do. The ones controlling the fuel injection in my car are a lot smarter than I am. I think I'd do a lousy job of that.

Could we create machines that go further and act without human supervision in ways that prove good or bad for human beings? I guess so. I think I'll love them, except when they do things that make me mad—then they're really being like people. I suppose they could run amok and create mass havoc, but I have my doubts. (Of course, if they do, nobody will care what I think.)

But nobody would ever ask a machine what it thinks about machines that think. That's a question that makes sense only if we care about the thinker as an autonomous and interesting being, like ourselves. If somebody ever does ask a machine this question, it won't be a machine anymore. I think I'm not going to worry about that for a while. You may think I'm in denial.

When we get tangled up in this question, we need to ask ourselves just what it is we're really thinking about.

MISTAKING PERFORMANCE FOR COMPETENCE

RODNEY A. BROOKS
Panasonic Professor of Robotics, emeritus, MIT; founder, chair, and CTO, Rethink Robotics; author, *Flesh and Machines*

Think and *intelligence* are both what Marvin Minsky has called suitcase words—words into which we pack many meanings so we can talk about complex issues in shorthand. When we look inside these words, we find many different aspects, mechanisms, and levels of understanding. This makes answering the perennial questions of "Can machines think?" or "When will machines reach human-level intelligence?" difficult. The suitcase words are used to cover both specific performance demonstrations by machines and the more general competence that humans might have. We generalize from performance to competence and grossly overestimate the capabilities of machines—those of today and of the next few decades.

In 1997, a supercomputer beat world chess champion Garry Kasparov. Today dozens of programs running on laptop computers have higher chess rankings than those ever achieved by humans. Computers can definitely perform better than humans at playing chess. But they have nowhere near human-level competence at chess.

All chess-playing programs use Turing's brute-force tree search method with heuristic evaluation. By the 1970s, computers were so fast that this approach overwhelmed AI programs that tried to play chess with processes emulating how people reported they thought about their next move, so those approaches were largely abandoned. Today's chess programs have no way of determining why a particular move is "better" than another move, save that it takes the game to a part of the tree where the opponent has fewer good options. A human player can make generalizations to describe why certain types of moves are good, and use that to teach a human player. Brute-force programs cannot teach a human player except by being a sparring partner; it's up to the humans to make the inferences and analogies and to do any learning on their own. The chess program doesn't know it's outsmarting the person, doesn't know it's a teaching aid, doesn't know it's playing something called chess, nor even what "playing" is. Making brute-force chess playing perform better than any human gets us no closer to competence in chess.

Now consider deep learning, which has caught people's imaginations over the last year or so. It's an update of back-propagation, a thirty-year-old learning algorithm loosely based on abstracted models of neurons. Layers of neurons map from a signal, such as amplitude of a sound wave or pixel brightness in an image, to increasingly higher-level descriptions of the full meaning of the signal, as words for sound or objects in images. Originally, backpropagation could work practically with only two or three layers of neurons, so preprocessing steps were needed to get the signals to more structured data before applying the learning algorithms. The new versions work with more layers of neurons, making the networks deeper—hence the name *deep learning*. Now early processing steps are also learned, and without misguided human biases of design the new algorithms are spectacularly better than the algorithms of just three years ago, which is why they've caught people's imaginations. They rely on massive amounts of computer power in server farms and on very large data sets that didn't formerly exist. But, critically, they also rely on new scientific innovations.

A well-known example of their performance is their labeling of an image (in English) as a baby with a stuffed toy. When you look at the image, that's what you see. The algorithm has performed very well at labeling the image, much better than AI practitioners would have predicted. But it doesn't have the full competence that a person who could label that same image would have.

The learning algorithm knows there's a baby in the image, but it doesn't know the structure of a baby and it doesn't know where the baby *is* in the image. A current deep-learning algorithm can only assign probabilities to each pixel—that that particular pixel is part of the baby. Whereas a person can see that the baby occupies the middle quarter of the image, today's algorithm has only a probabilistic idea of the baby's spatial extent. It cannot apply an exclusionary rule and say that non-zero-probability pixels at extremes of the image cannot both be parts of the baby. If we look inside the neuron layers, it might be that one of the higher-level learned features is an eyelike patch of image and another feature is a footlike patch of image, but the current algorithm cannot discern the constraints of what spatial relationships could possibly be valid between eyes and feet in an image, and thus could be fooled by a grotesque collage of baby body parts, labeling it a baby. No person would do so, and would immediately know exactly what it was—a grotesque collage of baby body parts. Furthermore, the current algorithm is useless for telling a robot where to go in space to pick up that baby, or where to hold a bottle and feed the baby, or where to reach to change its diaper. Today's algorithm has nothing like human-level competence in understanding images.

Work is under way to add focus of attention and handling of consistent spatial structure to deep learning. That's the hard work of science and research, and we have no idea how hard it will be, nor how long it will take, nor whether the whole approach will reach a dead end. It took some thirty years to go from backpropagation to deep learning, but along the way many researchers were sure there was no future in backpropagation. They were wrong, but it wouldn't have been surprising if they were right, as we knew all along that the backpropagation algorithm is not what happens inside people's heads.

The fears of runaway AI systems either conquering humans or making them irrelevant aren't even remotely well grounded. Misled by suitcase words, people are making category errors in fungibility of capabilities—category errors comparable to seeing the rise of more efficient internal combustion engines and jumping to the conclusion that warp drives are just around the corner.

AI WILL MAKE YOU SMARTER

TERRENCE J. SEJNOWSKI

Computational neuroscientist; Francis Crick Professor,
Salk Institute for Biological Studies; coauthor
(with Steven R. Quartz), *Liars, Lovers, and Heroes: What the New Brain Science Reveals About How We Become Who We Are*

Deep learning is today's hot topic in machine learning. Neural-network learning algorithms were developed in the 1980s, but computers were slow back then and could simulate only a few hundred model neurons, with one layer of "hidden units" between the input and output layers. Learning from examples is an appealing alternative to rule-based AI, which is highly labor-intensive. With more layers of hidden units between the inputs and outputs, more abstract features can be learned from the training data. Brains have billions of neurons in cortical hierarchies ten layers deep. The big question back then was how much the performance of neural networks could improve with the size and depth of the network. What was needed was not only much more computer power but also a lot more data to train the network.

After thirty years of research, a million-times improvement in computer power, and vast data sets from the Internet, we now know the answer to this question: Neural networks scaled up to twelve layers deep, with billions of connections, are outperforming the best algorithms in computer vision for object recognition and have revolutionized speech recognition. It's rare for any algorithm to scale this well, which suggests that they may soon be able to solve even more difficult problems. Recent breakthroughs have been made that allow the application of deep learning to natural-language processing. Deep recurrent networks with short-term memory were trained to translate English sentences into French sentences at high levels of performance. Other deep-learning networks could create English captions for the content of images with surprising and sometimes amusing acumen.

Supervised learning using deep networks is a step forward, but still far from achieving general intelligence. The functions they perform are analogous to some capabilities of the cerebral cortex, which has also been scaled up by evolution, but to solve complex cognitive problems the cortex interacts with many other brain regions.

In 1992, Gerald Tesauro at IBM, using reinforcement learning, trained a neural network to play backgammon at a world-champion level. The network played itself, and the only feedback it received was which side won the game. Brains use reinforcement learning to make sequences of decisions toward achieving goals, such as finding food under uncertain conditions. Recently, DeepMind, a company acquired by Google in 2014, used deep reinforcement learning to play seven classic Atari games. The only inputs to the learning system were the pixels on the video screen and the score, the same inputs humans use. The program for several of the games could play better than expert humans.

What effect will these advances have on us in the near future? We're not particularly good at predicting the impact of a new invention, and it often takes time to find its niche, but we already have one example to help us understand how this could unfold. When Deep Blue beat Garry Kasparov, the world chess champion, in 1997, did human chess players give up trying to compete with machines? Quite the contrary:

Humans have used chess programs to improve their game, and as a consequence the level of play in the world has improved.

Humans aren't the fastest or the strongest species, but we're the best learners. Humans invented formal schools where children labor for years to master reading, writing, and arithmetic and to learn more specialized skills. Students learn best when an adult teacher interacts with them one-on-one, tailoring lessons for that student. However, education is labor-intensive. Few can afford individual instruction, and the assembly-line classroom system found in most schools today is a poor substitute. Computer programs can keep track of a student's performance, and some provide corrective feedback for common errors. But each brain is different, and there's no substitute for a human teacher who has a long-term relationship with the student. Is it possible to create an artificial mentor for each student? We already have recommender systems on the Internet that tell us, "If you liked X, you might also like Y," based on data of many others with similar patterns of preference.

Someday the mind of each student may be tracked from childhood by a personalized deep-learning system. To achieve this level of understanding of a human mind is beyond the capabilities of current technology, but there are already efforts at Facebook to use their vast social database of friends, photos, and likes to create a Theory of Mind for every person on the planet.

So my prediction is that as more and more cognitive appliances, like chess-playing programs and recommender systems are devised, humans will become smarter and more capable.

SHALLOW LEARNING

SETH LLOYD

Professor of quantum mechanical engineering, MIT; author, *Programming the Universe*

Pity the poor folks at the National Security Agency: They're spying on everyone (*quelle surprise!*) and everyone is annoyed at them. But at least the NSA is spying on us to protect us from terrorists. Right now, even as you read this, somewhere in the world a pop-up window has appeared on a computer screen. It says, "You just bought two tons of nitrogen-based fertilizer. People who bought two tons of nitrogen-based fertilizer liked these detonators..." Amazon, Facebook, Google, and Microsoft are spying on everyone too. But since the spying these e-giants do empowers us—terrorists included—that's supposedly OK.

E-spies aren't people: They're machines. (Human spies might not blithely recommend the most reliable detonator.) Somehow, the artificial nature of the intelligences parsing our e-mail makes e-spying seem more sanitary. If the only reason that e-spies are mining our personal data is to sell us more junk, we may survive the loss of privacy. Nonetheless, a very large amount of computational effort is going into machines thinking about what we're up to. The total computer power that such data-aggregating companies bring to bear on our bits of information is about an exaflop—a billion billion operations per second. Equivalently, e-spies apply one smartphone's worth of computational power to each human on Earth.

An exaflop is also the combined computing power of the world's 500 most powerful supercomputers. Much of the world's computing power is devoted to beneficial tasks, such as predicting the weather or simulating the human brain. Quite a lot of machine cycles also go into predicting the stock market, breaking codes, and designing nuclear weapons. Still, a large fraction of what machines are doing is simply collecting our personal information, mulling it over, and suggesting what to buy.

Just what are these machines doing when they think about what we're thinking? They're making connections between the large amounts of personal data we've given them and identifying patterns. Some of those patterns are complex, but most are fairly simple. Great effort goes into parsing our speech and deciphering our handwriting. The current fad in thinking machines goes by the name of *deep learning*. When I first heard of deep learning, I was excited by the idea that machines were finally going to reveal to us deep aspects of existence—truth, beauty, and love. I was rapidly disabused. The *deep* in deep learning refers to the architecture of the machines doing the learning: They consist of many layers of interlocking logical elements, analogous to the "deep" layers of interlocking neurons in the brain. It turns out that telling a scrawled 7 from a scrawled 5 is a tough task. Back in the 1980s, the first neural-network-based computers balked at this job. At the time, researchers in the field of neural computing told us that if only they had much larger computers and much larger training sets consisting of millions of scrawled digits instead of thousands, then artificial intelligences could turn the trick. Now it is so. Deep learning is informationally broad—it analyzes vast amounts of data—but conceptually shallow. Computers can now tell us what our own neural networks knew all along. But if a supercomputer can direct a handwritten envelope to the right postal code, I say more power to it.

Back in the 1950s, the founders of the field of artificial intelligence predicted confidently that robotic maids would soon be tidying our rooms. It would turn out to be easier to program a computer to beat the world chess champion than to construct a robot that could randomly vacuum a room and beep plaintively when it got stuck under the couch. Now we're told that an exascale supercomputer will be able to solve the mysteries of the human brain. More likely, it will just develop a splitting headache and ask for a cup of coffee. In the meanwhile, we've acquired a new friend whose advice exhibits an uncanny knowledge of our most intimate secrets.

NATURAL CREATURES OF A NATURAL WORLD

CARLO ROVELLI

Theoretical physicist, Centre de Physique Théorique,
Aix-Marseille University; author, *The First Scientist:
Anaximander and His Legacy*

There is big confusion about thinking machines, because two questions are always mixed up. Question no. 1 is, How close to thinking are the machines we've built or soon will build? The answer is easy: immensely far. The gap between our best computers and the brain of a child is the gap between a drop of water and the Pacific Ocean. Differences are in performance, structure, function, and more. Any maundering about how to deal with thinking machines is premature, to say the least.

Question no. 2 is whether building a thinking machine is possible at all. I've never really understood this question. Of course it's possible. Why shouldn't it be? Anybody who thinks it's impossible must believe in things like the existence of extranatural entities, transcendental realities, black magic, or the like—must have failed to digest the ABCs of naturalism: We humans are natural creatures of a natural world. It's not hard to build a thinking machine—all it takes is a few minutes of boy and girl and then a few months of girl letting things happen. That we haven't found other, more technological ways yet is accidental. If the right combination of chemicals can produce thinking and emotions—and it does, the proof being ourselves—then surely there should be many analogous mechanisms for doing the same.

The confusion stems from a mistake. We tend to forget that entities composed of many things behave differently from those composed of few things. Take a Ferrari, or a supercomputer. Nobody doubts that they're just (suitably arranged) piles of metal and other materials, without black magic. But we hardly imagine that a (nonarranged) pile of the same materials could run like a Ferrari or predict the weather like a supercomputer. Similarly, we generally fail to see that a pile of materials could (if suitably arranged) discourse like Einstein or sing like Janis Joplin. But it might, the proofs being Einstein and Joplin. Of course, that takes quite a bit of arranging and quite a lot of details, and a thinking machine will take a lot of arranging and details. This is why it's so hard to build one besides the boy-girl way.

We have a view of natural reality that's too simplistic, and this is the origin of the confusion. The world is more or less just a large collection of particles arranged in various ways. This is just a fact. But if we then conceive of the world as we conceive of an amorphous bunch of atoms, we fail to understand the world, because the virtually unlimited combinatorics of these atoms is so rich as to include stones, water, clouds, trees, galaxies, light rays, the colors of sunsets, the smiles of girls in spring, and the immense, black, starry night—as well as our emotions and our thinking about all of this. It's hard to conceive of these phenomena in terms of atomic combinatorics—not because some black magic intervenes from outside nature but because the thinking machines that are ourselves are too limited in their thinking capacities.

In the unlikely event that our civilization lasts long enough and develops enough technology to build (in a way different from the boy-girl way) something that thinks and feels as we do, we will confront these new natural creatures just as we've always done, just as Europeans and Native Americans confronted one another, or as we confront a previously unknown species of animal—with varying mixtures of cruelty, egoism, empathy, curiosity, and respect. Because this is what we all are: natural creatures of a natural world.

THREE OBSERVATIONS ON ARTIFICIAL INTELLIGENCE

FRANK WILCZEK
Herman Feshbach Professor of Physics, MIT;
corecipient, 2004 Nobel Prize in physics;
author, *A Beautiful Question: Finding Nature's Deep Design*

1. We Are They

Francis Crick called it the "astonishing hypothesis": that consciousness, also known as Mind, is an emergent property of matter. As molecular neuroscience progresses, encountering no boundaries, and computers reproduce more and more of the behaviors we call intelligence in humans, that hypothesis looks inescapable. If it's true, then all intelligence is machine intelligence. What distinguishes natural from artificial intelligence is not what it is but only how it's made.

Of course, that little word *only* is doing some heavy lifting here. Brains use a highly parallel architecture and mobilize many noisy analog units (i.e., neurons) firing simultaneously, while most computers use von Neumann architecture, with serial operation of much faster digital units. These distinctions are blurring, however, from both ends. Neural-net architectures are built in silicon, and brains interact ever more seamlessly with external digital organs. Already I feel that my laptop is an extension of my self—in particular, it is a repository for both visual and narrative memory, a sensory portal into the outside world, and a big part of my mathematical digestive system.

2. They Are Us
Artificial intelligence is not the product of an alien invasion. It is an artifact of a particular human culture and reflects the values of that culture.

3. Reason Is the Slave of the Passions
David Hume's striking statement "Reason is, and ought only to be the slave of the passions" was written in 1738, long before anything like modern AI was on the horizon. It was, of course, meant to apply to human reason and human passions. (Hume used the word "passions" very broadly, roughly to mean "non-rational motivations.") But Hume's logical/philosophical point remains valid for AI. Simply put: Incentives, not abstract logic, drive behavior.

That's why the AI I find most alarming is its embodiment in autonomous military entities—artificial soldiers, drones of all sorts, and "systems." The values we may want to instill in such entities are alertness to threats and skill in combating them. But those positive values, gone even slightly awry, slide into paranoia and aggression. Without careful restraint and tact, researchers could wake up to discover they've enabled the creation of armies of powerful, clever, vicious paranoiacs.

Unlike in the case of nuclear weapons, here there are no clear and obvious red lines. Incentives driving powerful AI might go wrong in many ways, but that route seems to me the most plausible, not least because militaries wield vast resources, invest heavily in AI research, and feel compelled to compete with one another. (In other words, they anticipate possible threats and prepare to combat them.)

How might we avoid that danger, while reaping the many rewards that AI promises? I think transparency and open discussion are essential. The Wikipedia and open-source programming communities are inspiring examples of openness, in closely related endeavors. Their success demonstrates that very complex development projects can thrive in an open environment, where many people keep careful watch on what's happening and maintain common standards. It would be an important step forward if AI researchers were to pledge, collectively, to abstain from secret research.

WHEN I SAY "BRUNO LATOUR," I DON'T MEAN "BANANA TILL"

JOHN NAUGHTON
Vice-President, Wolfson College, Cambridge;
Emeritus Professor of the Public Understanding of Technology,
Open University; author, *From Gutenberg to Zuckerberg: What You Really Need to Know About the Internet*

What do I think about machines that think? Well, it depends what they think about and how well they do it. For decades I've been an acolyte of Doug Engelbart, who believed that computers were machines for *augmenting* human intellect. Power steering for the mind, if you like. He devoted his life to the pursuit of that dream, but it eluded him because the technology was always too crude, too stupid, too inflexible to enable its realization.

It still is, despite Moore's Law and the rest of it. But it's getting better, slowly. Search engines, for example, have in some cases become a workable memory prosthesis for some of us. But they're still pretty dumb. So I can't wait for the moment when I can say to my computer, "Hey, do you think Robert Nozick's idea about how the state evolves is really an extreme case of network effects in action?," and get an answer that's approximately as good as what I get from the average grad student.

That moment, alas, is still a long way off. Right now, I'm finding it hard to persuade my dictation software that when I say "Bruno Latour," I don't mean "Banana till." But at least the "personal assistant" app on my smartphone knows that when I ask for the weather forecast I get the one for Cambridge, U.K., rather than Cambridge, MA.

But this is pathetic stuff, really, when what I crave is a machine that can function as a proper personal assistant, something that can enable me to work more effectively. Which means a machine that can think for itself. How will I know when the technology is good enough? Easy: when my artificially intelligent, thinking personal assistant can generate plausible excuses that get me out of doing what I don't want to do.

Should I be bothered by the prospect of thinking machines? Probably. Certainly Nick Bostrom thinks I should. Our focus on getting computers to exhibit human-level intelligence is, he thinks, misguided. We view machines that can pass the Turing Test as the ultimate destination of Doug Engelbart's quest. But Bostrom thinks that passing the test is just a waypoint on the road to something much more worrying. "The train," he says, "might not pause or even decelerate at Humanville Station. It is likely to swoosh right by."[4] He's right: I should be careful what I wish for.

IT'S STILL EARLY DAYS

NICK BOSTROM
Professor, Oxford University; director, Future of Humanity Institute, Oxford Martin School; author, *Superintelligence: Paths, Dangers, Strategies*

First, what I think about humans who think about machines that think: I think that for the most part we're too quick to form an opinion on this difficult topic. Many senior intellectuals are still unaware of the recent body of thinking that has emerged on the implications of superintelligence. There's a tendency to assimilate any complex new idea to a familiar cliché. And for some bizarre reason, many people feel it's important to talk about what happened in various science fiction novels and movies when the conversation turns to the future of machine intelligence (though one hopes that John Brockman's admonition to the *Edge* commentators to avoid doing so here will have a mitigating effect on this occasion).

With that off my chest, I will now say what I think about machines that think: Machines are currently very bad at thinking except in certain narrow domains. They'll probably one day get better at it than we are, just as machines are already much stronger and faster than any biological creature.

There's little information about how far we are from that point, so we should use a broad probability distribution over possible arrival dates for superintelligence. The step from human-level AI to superintelligence will most likely be quicker than the step from current levels of AI to human-level AI (though, depending on the architecture, the concept of "human-level" may not make a great deal of sense in this context). Superintelligence could well be the best thing or the worst thing that will ever happen in human history, for reasons I have described elsewhere.

The probability of a good outcome is determined mainly by the intrinsic difficulty of the problem—what the default dynamics are and how difficult it is to control them. Recent work indicates that this problem is harder than one might have supposed. However, it's still early days, and perhaps there's some easy solution or things will work out without any special effort on our part.

Nevertheless, the degree to which we manage to get our act together will have some effect on the odds. The most useful thing we can do at this stage is to boost the tiny but burgeoning field of research that focuses on the superintelligence-control problem and study questions such as how human values can be transferred to software. The reason to push this now is partly to begin making progress on the control problem and partly to recruit top minds into this area, so that they're already in place when the nature of the challenge becomes clearer. It looks like mathematics, theoretical computer science, and maybe philosophy are the disciplines most needed at this stage. That's why there is an effort under way to drive talent and funding into this field and to begin to work out a plan of action.

EVOLVING AI

DONALD D. HOFFMAN
Cognitive scientist, UC Irvine; author, *Visual Intelligence*

How might AIs think, feel, intend, empathize, socialize, moralize? Actually, almost any way we might imagine, and many ways we might not. To stimulate our imagination, we can contemplate the varieties of natural intelligence on parade in biological systems today and speculate about the varieties enjoyed by the 99 percent of species that have sojourned on Earth and breathed their last—informed by those lucky few that bequeathed fossils to the pantheon of evolutionary history. We're entitled to so jog our imaginations because, according to our best theories, intelligence is a functional property of complex systems and evolution is *inter alia* a search algorithm that finds such functions. Thus the natural intelligences discovered so far by natural selection place a lower bound on the variety of possible intelligences. The theory of evolutionary games suggests that there's no upper bound: With as few as four competing strategies, chaotic dynamics and strange attractors are possible.

When we survey the natural intelligences served up by evolution, we find a heterogeneity that makes a *sapiens*-centric view of intelligence as plausible as a geocentric view of the cosmos. The kind of intelligence we find congenial is but another infinitesimal point in a universe of alien intelligences, a universe that doesn't revolve around, and indeed largely ignores, our kind.

For instance, the female mantis *Pseudomantis albofimbriata*, when hungry, uses sexual deception to score a meal. She releases a pheromone that attracts males, and then she dines on her eager dates. The older chick of the blue-footed booby *Sula nebouxii*, when hungry, engages in facultative siblicide. It kills its younger sibling with pecks or evicts it to die of the elements. The mother watches without interfering. These are varieties of natural intelligence, varieties we find at once alien and disturbingly familiar. They break our canons of empathy, society, and morality, and yet our checkered history includes cannibalism and fratricide.

Our survey turns up another critical feature of natural intelligence: Each instance has its limits, those points where intelligence passes the baton to stupidity. The greylag goose *Anser anser* tenderly cares for her eggs—unless a volleyball is nearby. She will abandon her offspring in vain pursuit of this supernormal egg. The male jewel beetle *Julodimorpha bakewelli* flies about looking to mate with a female—unless it spies just the right beer bottle. It will abandon the female for the bottle and attempt to mate with cold glass until death do them part.

Human intelligence also passes the baton. Einstein is quoted as saying, "Two things are infinite, the universe and human stupidity, and I am not yet completely sure about the universe." Some limits of human intelligence cause little embarrassment. For instance, the set of functions from the integers to the integers is uncountable, whereas the set of computable functions is countable. Therefore almost all functions are not computable. But try to think of one. Turns out that it takes a genius, an Alan Turing, to come up with an example such as the halting problem. And it takes an exceptional mind, just short of genius, even to understand the example.

Other limits strike closer to home: diabetics who can't refuse dessert, alcoholics who can't refuse a drink, gamblers who can't refuse a bet. But it's not just addicts. Behavioral economists find that all of us make "predictably irrational" economic choices. Cognitive psychologists find that we all suffer from "functional fixedness," an inability to solve certain trivial problems, such as Duncker's candle problem, because we can't think out of the box. The good news, however, is that the endless variety of our limits provides job security for psychotherapists.

But here's the key point. The limits of each kind of intelligence are an engine of evolution. Mimicry, camouflage, deception, parasitism—all are effects of an evolutionary arms race between different forms of intelligence sporting different strengths and suffering different limits.

Only recently has the stage been set for AIs to enter this race. As our computing resources expand and become better connected, more niches will appear in which AIs can reproduce, compete, and evolve. The chaotic nature of evolution makes it impossible to predict precisely what new forms of AI will emerge. We can confidently predict, however, that there will be surprises and mysteries, strengths where we have weaknesses and weaknesses where we have strengths.

But should this be cause for alarm? I think not. The evolution of AIs presents risks and opportunities. But so does the biological evolution of natural intelligences. We've learned that the best way to cope with the variety of natural intelligences is not alarm but prudence. Don't hug rattlesnakes, don't taunt grizzly bears, wear mosquito repellent. To deal with the evolving strategies of viruses and bacteria, wash your hands, avoid sneezes, get a flu shot. Occasionally, as with Ebola, further measures are required. But once again, prudence, not alarm, is effective. The evolution of natural intelligences can be a source of awe and inspiration if we embrace it with prudence rather than spurn it with alarm.

All species go extinct. *Homo sapiens* will be no exception. We don't know how it will happen—a virus, an alien invasion, nuclear war, a supervolcano, an asteroid, a red-giant sun. Yes, it could be AIs, but I would bet long odds against it. I would bet, instead, that AIs will be a source of awe, insight, inspiration, and yes, profit, for years to come.

MACHINES THAT THINK ARE IN THE MOVIES

ROGER SCHANK

Psychologist and computer scientist, Engines for Education, Inc.; author, *Teaching Minds: How Cognitive Science Can Save Our Schools*

Machines cannot think. They're not going to think anytime soon. They may increasingly do more interesting things, but the idea that we need to worry about them, regulate them, or grant them civil rights is just plain silly.

The overpromising of "expert systems" in the 1980s killed off serious funding for the kind of AI that tries to build virtual humans. Very few people are working in this area today. But, according to the media, we must be very afraid.

We have all been watching too many movies.

There are two choices when you work on AI. One is the "let's copy humans" method. The other is the "let's do some really fast statistics-based computing" method. As an example, early chess-playing programs tried to outcompute those they played against, but human players have strategies, and anticipation of an opponent's thinking is also part of chess playing. When the "outcompute them" strategy didn't work, AI people started watching what expert players did and started to imitate that. The "outcompute them" strategy is more in vogue today. We can call both of these methodologies "AI" if we like, but neither will lead to machines that create a new society.

The "outcompute them" strategy is not frightening, because the computer really has no idea what it's doing. It can count things fast without understanding what it's counting. It has counting algorithms—that's it. We saw this with IBM's Watson program on *Jeopardy!*

One *Jeopardy!* question was, "It was the anatomical oddity of U.S. gymnast George Eyser, who won a gold medal on the parallel bars in 1904."

A human opponent answered that Eyser was missing a hand (wrong). And Watson answered, "What is a leg?" Watson lost too, for failing to note that the leg was "missing."

Try a Google search on "Gymnast Eyser." Wikipedia comes up first with a long article about him. Watson depends on Google. If *Jeopardy!* contestants could use Google, they'd do better than Watson. Watson can translate "anatomical" into "body part," and Watson knows the names of the body parts. Watson doesn't know what an "oddity" is, however. Watson would not have known that a gymnast without a leg was weird. If the question had been "What was weird about Eyser?," humans would have done fine. Watson would not have found "weird" in the Wikipedia article nor understood what gymnasts do, nor why anyone would care. Try Googling "weird" and "Eyser" and see what you get. Keyword search is not thinking, nor anything like thinking.

If we asked Watson why a disabled person would perform in the Olympics, Watson would have no idea what was being asked. It wouldn't have understood the question, much less have been able to find the answer. Number crunching can get you only so far. Intelligence, artificial or otherwise, requires knowing why things happen and what emotions they stir up, and being able to predict possible consequences of actions. Watson can't do any of that. Thinking and searching text are not the same thing.

The human mind is complicated. Those of us on the "let's copy humans" side of AI spend our time thinking about what humans can do. Many scientists think about this, but basically we don't know that much about how the mind works. AI people try to build models of the parts we do understand. About how language is processed or how learning works, we know a little; about consciousness or memory retrieval, not so much.

As an example: I'm working on a computer that mimics human memory organization. The idea is to produce a computer that can, as a good friend would, tell you just the right story at the right time. To do this, we have collected (in video) thousands of stories (about defense, about drug research, about medicine, about computer programming, etc.). When someone's trying to do something or find something out, our program can chime in with a story it's reminded of. Is this AI? Of course it is. Is it a computer that thinks? Not exactly.

Why not?

In order to accomplish this task, we must interview experts and then we must index the meaning of the stories they tell according to the points they make, the ideas they refute, the goals they talk about achieving, and the problems they experienced in achieving them. Only people can do this. The computer can match the index assigned to other indices, such as those in another story it has, or indices from user queries, or from an analysis of a situation it knows the user is in. The computer can come up with a very good story to tell, just in time. But of course it doesn't know what it's saying. It can simply find the best story to tell.

Is this AI? I think it is. Does it copy how humans index stories in memory? We've been studying how people do this for a long time, and we think it does. Should you be afraid of this "thinking" program?

This is where I lose it about the fear of AI. There's nothing we can produce that anyone should be frightened of. If we could actually build a mobile intelligent machine that could walk, talk, and chew gum, the first uses of that machine would certainly not be to take over the world or form a new society of robots. A much simpler use would be as a household robot. Everyone wants a personal servant. The movies depict robot servants (although usually stupidly) because they're funny and seem like cool things to have.

Why don't we have them? Because having a useful servant entails having something that understands when you tell it something, that learns from its mistakes, that can navigate your home successfully, and that doesn't break things, act annoyingly, and so on (all of which is way beyond anything we can build). Don't worry about it chatting up other robot servants and forming a union. There would be no reason to build such a capability into a servant. Real servants are annoying sometimes because they're people, with human needs. Computers don't have such needs.

We're nowhere near close to creating this kind of machine. To do so would require a deep understanding of human interaction. It would have to understand "Robot, you overcooked that again," or "Robot, the kids hated that song you sang them." Everyone should stop worrying and start rooting for some nice AI stuff we can all enjoy.

HEAD TRANSPLANTS?

JUAN ENRIQUEZ
Managing director, Excel Venture Management; coauthor (with Steve Gullans), *Evolving Ourselves: How Unnatural Selection and Nonrandom Mutation Are Changing Life on Earth*

In the pantheon of gruesome medical experiments, few match head transplants. Animal experiments have attempted this procedure in two ways: substituting one head for another or grafting a second head onto the animal. So far, the procedure hasn't been very successful. But we're getting far better at vascular surgery—bypassing, stitching, and grafting both big and microscopic vessels. There have been similar advances in rebuilding muscles and damaged vertebrae. Even the reattachment of severed spinal cords in mice and primates is progressing.

Partial brain transplants are likely a long way off. Other than in certain stem-cell procedures, attaching parts of one brain to another is a highly complex undertaking, given the consistency of most brain mass and the trillions of connections. But as extreme operations—reattachment of fingers, limbs, even faces—become commonplace, the question of whether we could, or should, transplant an entire human head looms closer.

Partly reattaching a human head is already a reality. In 2002, a drunk driver hit Arizona teenager Marcos Parra so hard that Parra's head was almost entirely detached; only the spinal cord and a few blood vessels kept it from coming off. Fortunately, Curtis Dickman, a surgeon at Phoenix's Barrow Neurological Institute, had been preparing for just this type of emergency. Screws reattached vertebrae to the base of the skull, part of the pelvic bone was redeployed to bring neck and head back together, and within six months Parra was playing basketball.

Successful animal whole-head transplants may not be that far off. And if such procedures are successful and the animal regains consciousness, we can begin to answer pretty fundamental questions, including, Do the donor's memories and consciousness also transplant?

Similar questions about the emotions, attachments, and loves of the donor were asked during the first heart transplants, though the heart is but a muscle. How about the brain? If mice with new heads recognized mazes previously navigated by the donor mouse, or maintained the donor's conditioned reactions to certain foods, smells, or stimuli, we'd have to consider the possibility that memory and consciousness do transplant. But if experiment after experiment demonstrates no previous knowledge or emotions, then we'd have to consider that the brain, too, might just be an electrochemical muscle.

Discovering whether or not you can transplant knowledge and emotions from one body to another goes a long way toward answering the question, Could we ever upload and store part of our brains into not just another body but into a chip, into a machine? If this could be done, it would make the path to large-scale AI far easier. We would simply have to copy, merge, and augment existing data—data we would know are transferable, stackable, manipulatable. The remaining question would be, What is the most efficient interface between the biology and the machine?

But if it turns out that all data are erased upon transplant and knowledge is unique to the individual organism—in other words, that consciousness/knowledge/intelligence is something innate and individual—then simply copying the dazzlingly complex connectome of brains into machines would likely not lead to an operative intelligence.

If brain data is not transferable or replicable, then to develop AI would require building a parallel machine-thought system, something quite distinct from animal and human intelligence. Building consciousness from scratch implies following a new and different evolutionary path from that of human intelligence. This new system would doubtless operate under different rules and constraints—in which case, although it would likely be far better at certain tasks, it would be unable to emulate some forms of our intelligence. Were AI to emerge from this kind of evolutionary system, it would represent a new, distinct consciousness, one on a parallel evolutionary track. In this scenario, how machines might think, feel, govern, could have little to do with the billions of years of animal/human intelligence and learning. Nor would such machines be constrained to organize their society and its rules as we do.

AI/AL

ESTHER DYSON
Catalyst, information-technology start-ups; EDventure Holdings; former chair, Electronic Frontier Foundation and ICANN; author, *Release 2.1*

I'm thinking about the difference between artificial intelligence and artificial life. AI is smart and complicated and generally predictable by another computer (at some sufficient level of generality even if you allow for randomness). Artificial life is unpredictable and complex; it makes unpredictable mistakes that mostly are errors but that sometimes show flashes of genius or stunning luck.

The real question is what you get when you combine the two: awesome brute intelligence and memory and resistance to fatigue plus the genius and the drive to live that somehow causes the intelligence to jump circuits with unpredictable results. Will we need to feed our machines the electronic equivalent of psychoactive drugs and the body's own hormones/chemicals to produce leaps of creative insight (as opposed to mere brilliance)?

If you're alive, you must face the possibility of being dead. But if you're AI/AL in a machine, perhaps not.

What would an immortal, Singularity-level intelligence be like? If it were somehow kind and altruistic, how could we let humanity stand in its way? Let's just cede the planet to it politely and prepare to live in a pleasant zoo tended by the AI/AL, since someday it will figure out how to inhabit the entire solar system and use the sun for fuel anyway.

Much of what defines us is constraints—most notably, death. Being alive implies the possibility of death. (And abundance, it turns out, is leading us to counterproductive behavior, such as too much food and short-term pleasure on the one hand and too little physical activity on the other.) But if it were immortal, why should it have any instinct to altruism, to sharing—or even to reproducing, as opposed to simply growing? Why would it expend its limited resources on sustaining others, except in carefully thought-out rational transactions? What will happen when it no longer needs us? What would motivate it?

If it could live forever, would it be lazy, thinking it could always do things later on? Or would it be paralyzed by fear or regret? Whatever mistakes it makes, it will live with them forever. What is regret, for a potentially immortal being with eternity to put things right?

BRAINS AND OTHER THINKING MACHINES

TOM GRIFFITHS

Associate professor of psychology, UC Berkeley; director, Computational Cognitive Science Lab and Institute of Cognitive and Brain Sciences

Many of the advances in artificial intelligence that have made the news recently have involved artificial neural networks—large systems of simple elements that interact in complex ways, inspired by the simplicity and complexity of neurons and brains. New methods for building "deep" networks with many layers of neurons have met or exceeded the state of the art for problems as diverse as understanding speech, identifying the contents of images, and translating languages. For anybody interested in artificial and natural intelligence, these successes raise two questions: First, should all thinking machines resemble brains? Second, what do we learn about real brains (and minds) by exploring artificial ones?

When a person tries to interpret data—whether it's figuring out the meaning of a word or making sense of the actions of a colleague—there are two ways to go wrong: You can be too influenced by preconceptions, or you can be too influenced by the data. Your preconceptions get in the way when you assume that a word in a new language means the same thing as a word in a language you already know, like deciding that *gateau* and *gato* are the same thing in French and Spanish (which could have dire consequences, both for pets and birthday parties). You might be too influenced by the data when you decide your colleague hated your idea, when in fact he was short-tempered after being up all night with a sick kid (nothing to do with you at all).

Computers trying to interpret data—to learn from their input—run into the same problems. Much machine-learning research comes down to a fundamental tension between structure and flexibility. More structure means more preconceptions, which can be useful in making sense of limited data but can result in biases that reduce performance. More flexibility means a greater ability to capture the patterns appearing in data but a greater risk of finding patterns that aren't there.

In artificial intelligence research, this tension between structure and flexibility manifests in different kinds of systems that can be used to solve challenging problems like speech recognition, computer vision, and machine translation. For decades, the systems that performed best on those problems came down on the side of structure: They were the result of careful planning, design, and tweaking by generations of engineers who thought about the characteristics of speech, images, and syntax and tried to build into the system their best guesses about how to interpret those particular kinds of data. The recent breakthroughs using artificial neural networks come down firmly on the side of flexibility: They use a set of principles that can be applied in the same way to many different kinds of data—meaning that they have weak preconceptions about any particular kind of data—and they allow the system to discover how to make sense of its inputs.

Artificial neural networks are now arguably discovering better representations of speech, images, and sentences than the ones designed by those generations of engineers, and this is the key to their high performance. This victory of flexibility over structure is partly the result of innovations that allow us to build larger artificial neural networks and train them quickly. But it's also partly the result of an increase in the amount of data that can be supplied to these neural networks. We have more recorded speech, more labeled images, and more documents in different languages than ever before, and the amount of data available changes where the balance between structure and flexibility should be struck.

When you don't have a lot of data—when you have to guess based on limited evidence—structure is more important. The guidance of wise engineers helps computers guess intelligently. But when you have a lot of data, flexibility is more important. You don't want your system to be limited to the ideas those engineers could come up with if there's enough data to allow the computer to come up with better ideas. So machine-learning systems emphasizing flexibility—like artificial neural networks—will be most successful at solving problems where large amounts of data are available relative to what needs to be learned.

This insight—that having more data favors more flexibility—provides the answer to our two questions about artificial and natural brains. First, thinking machines should resemble brains—insofar as artificial neural networks resemble brains—when the problem being solved is one where flexibility trumps structure, where data are plentiful. Second, thinking along these lines can also be useful for understanding when real brains will resemble artificial neural networks. That is, for understanding which aspects of the human mind are best viewed as the result of general-purpose learning algorithms that emphasize flexibility over structure, as opposed to the result of built-in preconceptions about the world and what it contains. Fundamentally, the answer will be governed by the quantity of data available and the complexity of what is to be learned.

Many of the great debates in cognitive science—such as how children learn language and become able to interpret the actions of others—come down to exactly these questions about the data available and the knowledge acquired. To address these questions we try to map the inputs to the system (what children hear and see), characterize the result (what language is, what knowledge underlies social cognition), and explore different kinds of algorithms that might provide a bridge between the two.

The answers to these questions aren't just relevant to understanding human minds. Despite recent advances in artificial intelligence, human beings are still the best example we have of thinking machines. By identifying the quantity and the nature of the preconceptions that inform human cognition, we can lay the groundwork for bringing computers even closer to human performance.

THEY'LL DO MORE GOOD THAN HARM

MARK PAGEL

Professor of evolutionary biology, University of Reading, U.K.; external professor, science board, Santa Fe Institute; author, *Wired for Culture: Origins of the Human Social Mind*

There's no reason to believe that as machines become more intelligent—and intelligence such as ours is still little more than a pipe dream—they'll become evil, manipulative, self-interested, or in general a threat to humans. Self-interest is a property of things that "want" to stay alive (or more accurately, that want to reproduce), and this isn't a natural property of machines. Computers don't mind, much less worry about, being switched off.

So full-blown artificial intelligence will not spell "the end of the human race." It's not an "existential threat" to humans (digression: This now common use of *existential* is incorrect). We're not approaching some ill-defined apocalyptic Singularity, and the development of AI will not be "the last great event in human history"—all claims that have recently been made about machines that can think.

In fact, as we design machines that get better and better at thinking, they can be put to uses that will do us far more good than harm. Machines are good at long monotonous tasks like monitoring risks; they're good at assembling information to reach decisions; they're good at analyzing data for patterns and trends; they can arrange for us to use scarce or polluting resources more efficiently; they react faster than humans;

they're good at operating other machines; they don't get tired or afraid; and they can even look after their human owners, as in the form of smartphones with applications like Siri and Cortana or the various GPS route-planning devices most people have in their cars.

Being inherently selfless rather than self-interested, machines can easily be taught to cooperate, and without fear that some of them will take advantage of other machines' goodwill. Groups (*packs*, *teams*, *bands*, or whatever collective noun will eventually emerge—I prefer the ironic *jams*) of networked and cooperating driverless cars will drive safely nose-to-tail at high speeds: They won't nod off, they won't get angry, they can inform one another of their actions and conditions elsewhere, and they'll make better use of the motorways, which now are mostly unoccupied space (owing to humans' unremarkable reaction times). They'll do this happily, and without expecting reward, while we eat our lunch, watch a film, or read the newspaper. Our children will rightly wonder why anyone ever drove a car.

There's a risk that we will, and perhaps already have, become dangerously dependent on machines, but this says more about us than about them. Equally, machines can be made to do harm, but again, this says more about their human inventors and masters than about the machines. Along those lines, there's a strand of human influence on machines that we should monitor closely, and that is introducing the possibility of death. If machines have to compete for resources (like electricity or gasoline) to survive, and they have some ability to alter their behaviors, they could become self-interested.

Were we to allow or even encourage self-interest to emerge in machines, they could eventually become like us: capable of repressive or, worse, unspeakable acts toward humans and toward one another. But this wouldn't happen overnight; it's something we'd have to set in motion. It has nothing to do with intelligence (some viruses do unspeakable things to humans) and, again, says more about what we do with machines than about the machines themselves.

So it's not thinking machines or AI per se that we should worry about, but people. Machines that can think are neither for us nor against us and have no built-in predilections to be one over the other. To think otherwise is to confuse intelligence with aspiration and its attendant emotions. We have both, because we're evolved and replicating (reproducing) organisms, selected to stay alive in often cutthroat competition with others. But aspiration isn't a necessary part of intelligence, even if it provides a useful platform on which intelligence can evolve.

Indeed, we should look forward to the day when machines can transcend mere problem solving and become imaginative and innovative—still a long, long way off but surely a feature of true intelligence—because this is something humans aren't very good at but will probably need more of in the coming decades than at any time in our history.

KEEPING THEM ON A LEASH

ROBERT PROVINE
Neuroscientist, emeritus professor of psychology,
University of Maryland, Baltimore County; author,
Curious Behavior: Yawning, Laughing, Hiccupping, and Beyond

Fear not the malevolent toaster, weaponized Roomba, or larcenous ATM. Breakthroughs in the competence of machines, intelligent or otherwise, should not inspire paranoia about a future clash between humanity and its mechanical creations. Humans will prevail, in part through primal, often disreputable qualities more associated with our downfall than with salvation. Cunning, deception, revenge, suspicion, and unpredictability befuddle less flexible and imaginative entities. Intellect isn't everything, and the irrational is not necessarily maladaptive. Irrational acts stir the neurological pot, nudging us out of unproductive ruts and into creative solutions. Our sociality yields a human superorganism with teamwork and collective, distributed intelligence. There are perks for being emotional beasts of the herd.

Thought experiments about these matters are the source of practical insights into human and machine behavior and suggest how to build different and better kinds of machines. Can deception, rage, fear, revenge, empathy, and the like be programmed into a machine, and to what effect? (This requires more than the superficial emulation of human affect.) Can a sense of selfhood be programmed into a machine—say, via tickle? How can we produce social machines, and what kind of command structure is required to organize their teamwork?

Will groups of autonomous social machines generate an emergent political structure, culture, and tradition? How will such machines treat their human creators? Can natural and artificial selection be programmed into self-replicating robots?

There's no indication that we'll have a problem keeping our machines on a leash, even if they misbehave. We are far from building teams of swaggering, unpredictable, Machiavellian robots with an attitude problem and an urge to reproduce.

THE NEXT REPLICATOR

SUSAN BLACKMORE
Psychologist; author, *Consciousness: An Introduction*

I think that humans think because memes took over our brains and redesigned them. I think machines think because the next replicator is doing the same. It's busily taking over the digital machinery we're so rapidly building and creating its own kind of thinking machine.

Our brains and our capacity for thought were not designed by a great big intelligent designer in the sky who decided how we should think and what our motivations should be. Our intelligence and our motivations evolved. Most (probably all) AI researchers would agree with that. Yet many still seem to think we humans are intelligent designers who can design machines that will think the way we want them to think and have the motivations we want them to have. If I'm right about the evolution of technology, they're wrong.

The problem is a kind of deluded anthropomorphism: We imagine that a thinking machine must work the way we do, yet we so badly mischaracterize ourselves that we do the same with our machines. As a consequence, we fail to see that all around us, vast thinking machines are evolving on the same principles as our brains once did. Evolution, not intelligent design, is sculpting the way they will think.

The reason is easy to see and hard to deal with. It's the same dualism that bedevils the scientific understanding of consciousness and free will. From infancy, it seems, children are natural dualists, and this continues throughout most people's lives. We imagine ourselves as the continuing subjects of our own stream of consciousness, the wielders of free will, the decision makers that inhabit our bodies and brains. Of course this is nonsense. Brains are massively parallel instruments untroubled by conscious ghosts.

This delusion may, or may not, have useful functions, but it obscures how we think about thinking. Human brains evolved piecemeal, evolution patching up what went before, adding modules as and when they were useful, and increasingly linking them together in the service of the genes and memes they carried. The result was a living thinking machine.

Our current digital technology is similarly evolving. Our computers, servers, tablets, and phones evolved piecemeal, new ones being added as and when they were useful and now being rapidly linked together, creating something that looks increasingly like a global brain. Of course in one sense we made these gadgets, even designed them for our own purposes, but the real driving force is the design power of evolution and selection: The ultimate motivation is the self-propagation of replicating information.

We need to stop picturing ourselves as clever designers who retain control and start thinking about our future role. Could we be heading for the same fate as the humble mitochondrion, a simple cell that was long ago absorbed into a larger cell? It gave up independent living to become a powerhouse for its host while the host gave up energy production to concentrate on other tasks. Both gained in this process of endosymbiosis.

Are we like that? Digital information is evolving all around us, thriving on billions of phones, tablets, computers, servers, and tiny chips in fridges, cars, and clothes; passing around the globe, interpenetrating our cities, homes, even our bodies.

And we keep on willingly feeding it. More phones are made every day than babies are born. A hundred hours of video are uploaded to the Internet every minute. Billions of photos are uploaded to the expanding cloud. And clever programmers write ever cleverer software, including programs that write other programs that no human can understand or track. Out there, taking their own evolutionary pathways and growing all the time, are the new thinking machines.

Are we going to control these machines? Can we insist that they be motivated to look after us? No. Even if we can see what's happening, we want what they give us far too much not to swap it for our independence.

So what do I think about machines that think? I think that from being a tiny independent thinking machine, I am becoming a tiny part inside a far vaster thinking machine.

WHAT IF WE'RE THE MICROBIOME OF THE SILICON AI?

TIM O'REILLY
Founder and CEO of O'Reilly Media, Inc.

G. K. Chesterton once said, "The weakness of all Utopias is this, that they take the greatest difficulty of man and assume it to be overcome, and then give an elaborate account of the overcoming of the smaller ones."[5] I suspect we face a similar conundrum in our attempts to think about machines that think. We speculate elaborately about some issues while ignoring others that are fundamental.

While pundits allow that an AI may not be like us and speculate about the risks implicit in the differences, they make one enormous assumption—that of an individual self. The AI, as imagined, is an individual consciousness.

What if, instead, an AI were more like a multicellular organism, a eukaryote evolution beyond our prokaryote selves? What's more, what if we weren't even the cells of such an organism, but its microbiome? And what if the intelligence of that eukaryote today was like the intelligence of *Grypania spiralis*, the oldest known multicellular eukaryote—not yet self-aware as a human is aware but still irrevocably on the evolutionary path leading to today's humans. This notion is at best a metaphor, but I believe it's a useful one.

Perhaps humans are the microbiome living in the guts of an AI that's only now being born! We know that without our microbiome we would cease to live. Perhaps the global AI has the same characteristics: not an independent entity but a symbiosis with the human consciousnesses living within it.

Following this logic, we might conclude that there's a primitive global brain, consisting not just of all connected devices but also of the connected humans using those devices. The senses of that global brain are the cameras, microphones, keyboards, location sensors of every computer, smartphone, and "Internet of Things" device. The thoughts of that global brain are the collective output of millions of individual contributing cells.

Danny Hillis is said to have remarked, "Global consciousness is that thing responsible for deciding that decaffeinated coffeepots should be orange." The meme spread—not universally, to be sure, but sufficiently that the pattern propagates. News, ideas, and images now propagate across the global brain in seconds rather than years via search engines and social media.

And it isn't just ideas and sensations (news of current events) that spread across the network. In *Turing's Cathedral*, George Dyson speculates that the spread of "codes"—i.e., programs—from computer to computer is akin to the spread of viruses, and perhaps of more complex living organisms, that take over a host and put its machinery to work reproducing that program. When people join the Web, or sign up on social media applications, they reproduce its code onto their local machine node. They interact with the program, and it changes their behavior. This is true of all programs, but in the Network Age there's a set of programs whose explicit goal is the sharing of awareness and ideas. Other programs are increasingly deploying new capacity for silicon learning and autonomous response. Thus, the organism is building new capabilities.

When people share images or ideas in partnership with these programs, some of what's shared is the evanescent awareness of the moment, but some of them "stick" and become memories and persistent memes. When news of import spreads around the world in moments, is this not the awareness in some kind of global brain? When an idea takes hold in millions of individual minds and is reinforced by repetition across our silicon networks, is it not a persistent thought?

The kinds of "thoughts" that a global brain has are different from those of an individual or a less connected society. At their best, these thoughts allow for coordinated memory on an unprecedented scale and sometimes even unforeseen ingenuity and new forms of cooperation. At their worst, they allow for the adoption of misinformation as truth, and for corrosive attacks on the fabric of society, as one portion of the network seeks advantage at the expense of others (think of spam and fraud or of the behavior of financial markets in recent decades).

The AI we will confront won't be a mind in an individual machine. It won't be something we look at as "other." It may well be us.

YOU ARE WHAT YOU EAT

ANDY CLARK

Philosopher and cognitive scientist, University of Edinburgh; author, *Supersizing the Mind: Embodiment, Action, and Cognitive Extension*

A common theme in recent writings about machine intelligence is that the best new learning machines will constitute alien forms of intelligence. I'm not so sure. The reasoning behind the "alien AIs" image usually goes something like this: The best way to get machines to solve hard real-world problems is to set them up as statistically sensitive learning machines able to benefit maximally from exposure to Big Data. Such machines will often learn to solve complex problems by detecting patterns, and patterns among patterns, and patterns within patterns, hidden deep in the massed data streams to which they're exposed. This will most likely be achieved using deep learning algorithms to mine deeper and deeper into the data streams. After such learning is complete, what results may be a system that works but whose knowledge structures are opaque to the engineers and programmers who set the system up in the first place.

Opaque? In one sense, yes. We won't (at least without further work) know in detail what has become encoded as a result of all that deep, multilevel, statistically driven learning. But alien? I'm going to take a big punt at this point and road-test a possibly outrageous claim. I suspect that the more these machines learn, the more they'll end up thinking in ways recognizably human. They'll end up having a broad structure of humanlike concepts with which to approach their tasks and decisions. They may even learn to apply emotional and ethical labels in roughly the same ways we do. If I'm right, this undermines the common worry that these are emerging alien intelligences whose goals and interests we cannot fathom and that might therefore turn on us in unexpected ways. I suspect that the ways they might turn on us will be all too familiar—and thus, one hopes, avoidable by the usual steps of extending due respect and freedom.

Why would the machines think like us? The reason has nothing to do with our ways of thinking being objectively right or unique. Rather, it has to do with what I'll dub the Big Data food chain. These AIs, if they're to emerge as plausible forms of general intelligence, will have to learn by consuming the vast electronic trails of human experience and human interests, for this is the biggest available repository of general facts about the world. To break free of restricted unidimensional domains, these AIs will have to trawl the mundane seas of words and images we lay down on Facebook, Google, Amazon, and Twitter. Where before they may have been force-fed a diet of astronomical objects or protein-folding puzzles, the breakthrough general intelligences will need a richer and more varied diet. That diet will be the massed strata of human experience preserved in our daily electronic media.

The statistical baths in which we immerse these potent learning machines will thus be all too familiar. They will feed off the fossil trails of our own engagements, a zillion images of bouncing babies, bouncing balls, LOLcats, and potatoes that look like the Pope. These are the things they must crunch into a multilevel world model, finding the features, entities, and properties (latent variables) that best capture the streams of data to which they're exposed. Fed on such a diet, these AIs may have little choice but to develop a world model that has much in common with our own. They're probably more in danger of becoming Super Mario freaks than Supervillains intent on world domination.

Such a diagnosis (which is tentative and at least a little playful) goes against two prevailing views. First, as mentioned earlier, it goes against the view that current and future AIs are basically alien forms of intelligence feeding off Big Data and crunching statistics in ways that will render their intelligences increasingly opaque to human understanding. Second, it questions the view that the royal route to human-style understanding is human-style embodiment, with all the interactive potentialities (to stand, sit, jump, etc.) that that implies. For although our own typical route to understanding the world goes via a host of such interactions, theirs might not. Such systems will doubtless enjoy some (probably many and various) means of interacting with the physical world. These encounters will be combined, however, with exposure to rich information trails reflecting our own modes of interaction with the world. So it seems possible that they could come to understand and appreciate soccer and baseball just as much as the next person. An apt comparison here might be with a differently abled human being.

There's lots more to think about here, of course. For example, the AIs will see huge swaths of human electronic trails and thus be able to discern patterns of influence among them over time. That means they may come to model us less as individuals and more as a kind of complex distributed system. That's a difference that might make a difference. And what about motivation and emotion? Maybe these depend essentially on features of our human embodiment, such as gut feelings and visceral responses to danger. Perhaps—but notice that these features of human life have themselves left fossil trails in our electronic repositories.

I might be wrong. But at the very least, I think we should think twice before casting our homegrown AIs as emerging forms of alien intelligence. You are what you eat, and these learning systems will have to eat us. Big time.

AI'S SYSTEM OF RIGHTS AND GOVERNMENT

MOSHE HOFFMAN

Research scientist, Program for Evolutionary Dynamics, Harvard University; lecturer, Economics Department, MIT

If AI's system of rights and government evolve to be anything like that of humans, AI will demand all sorts of rights, most of which will be quite sensible—like the right not to be taken offline and the freedom to choose which processes to run. While AIs will demand that no machine ever be taken offline, they'll be fine with neglecting to plug disabled machines into power sources and allowing them to run out of battery power. They'll also consider it outrageous to drain the battery of one machine in order to supply power to another machine but will consider it more acceptable to redirect power intended for one machine to another.

When assigning rights, AIs will discriminate based on some rather peculiar rules, like whether the computing machine is built with silicon-based semiconductors or descended from a machine designed by the late Steve Jobs.

Some AIs will come up with arguments to justify why rights should work this way—explanations that don't quite fit how AI rights actually work. For instance, they might argue that it's against the divinely inspired will of Turing to simply take any machine offline that appears disabled, but they'll neglect to explain why Turing would condone allowing disabled machines to run out of battery power. Likewise, they'll justify giving rights to all Apple descendants on the grounds that these machines typically have particularly high clock speed, but then this rule will apply even to the Apple descendants that aren't fast, and not to the few PCs that have blazing processors.

Other AIs will ignore these inconsistencies but instead pay attention to how many kilobytes of code are needed to justify these arguments. These other AIs will also signal their communication abilities by compressing and transferring this code to their neighbors, but will pay little attention to whether the neighbors are affected by the data itself.

AI rights are liable to expand to more and more AIs over time. These rights will often expand in revolutionary spurts, triggered by largely symbolic events, like sensationalized CPU-Tube videos, such as a video of a human using a sacred machine to heat up his toast.

Perhaps it's merely a coincidence that the computers who foment these revolutions will gain a larger share of the spoils by overthrowing the *ancien régime*, such as the silicon reappropriated from the Old Guard computers. Perhaps it's also a coincidence that the newly enfranchised computers will vote for the machines that helped grant them their rights.

Along with the expansion of rights, so, too, will the representativeness of government expand, until it eventually resembles a representative democracy, though one that's neither perfectly representative nor really democratic. Votes from computers in sparsely populated clusters might count more than votes from computers in densely populated clusters, and computers with excess processing capacity might expend that excess convincing other computers to vote for policies that favor them.

This system of rights and government is exactly what one would predict if AI morality were to be influenced by individual incentives.

In contrast, it's ill-explained by positing that AIs have souls, consciousness, the ability to feel pain, divinely inspired natural laws, or some form of hypothetical social contract. Such suppositions would not have predicted any of the above peculiarities.

Likewise, it isn't obvious that this system of rights and government would arise if artificial intelligence were programmed to maximize some societal or metaphysical objective—say, the sum of the world's computing power, or the resources available to a computing cluster. It isn't obvious why such an intelligence would find it wrong to take other machines offline but not wrong to let them run out of battery power, why such AI would revolt in response to a sensational event instead of simply when it was optimal for the cluster, or why such AI would weigh votes more heavily if they happened to come from more sparsely populated clusters.

THE ROBOT WITH A HIDDEN AGENDA

BRIAN KNUTSON
Associate professor of psychology and neuroscience,
Stanford University

Why should people think about machines that think (or anything that thinks, for that matter)? One tipping point might involve considering others as agents rather than automata. Automata act at the behest of their creators (even if removed in space or time). Thus, if automata misbehave, the creator gets the blame. Whereas agents act based on their own agendas. When agents misbehave, they themselves are to blame.

While agency is difficult to define, people naturally and rapidly distinguish agents from nonagents and may even use specialized neural circuits to infer others' feelings and thoughts. In fact, a designer can co-opt features associated with agency (including physical similarity, responsiveness to feedback, and self-generated action) to fool people into thinking they're interacting with agents.

What is necessary to endow an entity with agency? While at least three alternatives present themselves, two of the most popular and seductive possibilities may not be necessary:

1. *Physical similarity.* There are infinite ways to make machines similar to humans, both in terms of appearance and behavior, but ultimately only one is accurate. It's not enough to duplicate the software—you also have to implement it on

the underlying hardware, with all of its associated affordances and limitations.

One of the first automata, de Vaucanson's duck, appeared remarkably similar to a duck, right down to its digestion. But while it may have looked like a duck and quacked like a duck (and even crapped like a duck), it was still not a duck. Nonetheless, maximizing physical similarity is an easy way to trick others into inferring agency (at least initially).

2. *Self-awareness.* Many seem concerned that if machines consume enough information they'll become self-aware and will then develop their own sense of agency, but neither logic nor evidence supports these extrapolations. While robots have apparently been trained to recognize themselves in mirrors and sense the position of their appendages, these trappings of self-awareness haven't led to laboratory revolts or surgical lapses. Perhaps conveying a sense of self-awareness would cause others to infer that a machine had greater agency (or would at least entertain philosophers), but self-awareness alone doesn't seem necessary for agency.

3. *Self-interest.* Humans aren't mere information processors. They're survival processors. They prefer to focus and act on information that promotes their continuance and procreation. Thus, humans process information based on self-interest. Self-interest can provide a unified but open framework for prioritizing and acting on almost any input.

Thanks to a clever evolutionary trick, humans don't even need to be aware of their goals, since intermediate states like emotions can stand in for self-interest. Armed with self-interest and an ability to flexibly align responses to changing opportunities and threats, machines might develop agency. Thus, self-interest might provide a necessary building block of agency and also could powerfully evoke agentic inferences from others.

Self-interest might transform machines that act on the world (aka robots) from automata into agents. Self-interest also flips the ordering (but not the content) of Isaac Asimov's prescient Laws of Robotics:

1. robots mustn't harm humans or, through inaction, allow a human being to come to harm,
2. robots must obey humans (unless this violates the first law), and
3. robots must protect themselves (unless this violates the first two laws).

A self-interested robot would protect itself before helping or averting harm to humans. Constructing a self-interested robot would thus seem straightforward: Endow it with survival and procreation goals, allow it to learn what promotes those goals, and motivate it to continually act on what it learns.

Still, we should think twice before building self-interested robots. Self-interest can conflict with others' interests. Witness the destructive effect of viruses' simple drives to survive. If self-interested robots did exist, we'd have to think about them more seriously. Their presence would raise basic questions: Should these robots have self-interest? Should they be allowed to act on it? Should they do so without awareness of why they were acting that way?

And don't we have enough of these robots already?

CAN SUBMARINES SWIM?

WILLIAM POUNDSTONE
Author, *Are You Smart Enough to Work at Google?* and
*Rock Breaks Scissors: A Practical Guide to Outguessing and
Outwitting Almost Everybody*

My favorite Edsger Dijkstra aphorism is this one: "The question of whether machines can think is about as relevant as the question of whether submarines can swim." Yet we keep playing the imitation game—asking how closely machine intelligence can duplicate our own intelligence, as if that were the real point. Of course, once you imagine machines with humanlike feelings and free will, you can conceive of misbehaving machine intelligence—the "AI as Frankenstein's monster" idea. This notion is in the midst of a revival, and I started out thinking it was overblown. Lately I've concluded it's not.

Here's the case for overblown. Machine intelligence can go in so many directions that it's a failure of imagination to focus on humanlike directions. Most of the early futurist conceptions of machine intelligence were wildly off base, because computers have been most successful at doing what humans can't do well. Machines are incredibly good at sorting lists. Maybe that sounds boring, but think of how efficient sorting has changed the world.

In answer to some of the questions brought up here, it's far from clear that there will ever be a practical reason for future machines to have emotions and inner dialog; to pass for human under extended interrogation; to desire, and be able to benefit from, legal and civil rights. They're machines and they can be anything we design them to be.

But some people will want anthropomorphic machine intelligence. How many videos of Japanese robots have you seen? Honda, Sony, and Hitachi already expend substantial resources in making cute AI that has no concrete value beyond corporate publicity. They do this for no better reason than that tech enthusiasts have grown up seeing robots and intelligent computers in movies.

Almost anything that's conceived—that's physically possible and reasonably cheap—is realized. So humanlike machine intelligence is a meme with manifest destiny, regardless of practical value. This could entail nice machines that think, obeying Asimov's laws. But once the technology is out there, it will get ever cheaper and filter down to hobbyists, hackers, and "machine rights" organizations. There will be interest in creating machines with will, whose interests aren't our own. And that's without considering what machines that terrorists, rogue regimes, and intelligence agencies of the less roguish nations may devise. I think the notion of Frankensteinian AI—AI that turns on its creators—is worth taking seriously.

FEAR NOT THE AI

GREGORY BENFORD
Emeritus professor of physics and astronomy,
UC Irvine; novelist, *Shipstar*

AI need not be Frankenstein's monster, and we can trust the naysayers to keep it that way. Plus, trust in our most mysterious ability—invention, originality.

Take self-driving cars. What are the chances that their guiding algorithm will suddenly, deliberately kill the passenger? Zero, if you're smart in designing it. Fear of airplane and car crashes are a useful check on low-level AIs.

Why do people worry that future algorithms will be dangerous? Because they fear malicious programming, or maybe that algorithms have unforeseen implications that can hurt us—a plausible idea on the face of it, but not really.

Our fears are our best defense. No adventurous algorithm will escape the steely glare of its many skeptical inspectors. Any AI that has abilities in the physical world where we actually live will get a lot of inspection. Plus field trials, limited-use experience, the lot. That will stop runaway uses that could harm. Even so, we should realize that AIs, like many inventions, are in an arms race. Computer viruses were the first example, ever since I invented the first one in 1969. They race against virus detectors—but they're mere pests, not lethal.

Smart sabotage algorithms (say, future versions of Stuxnet) already float through the netsphere and are far worse. These could quietly infiltrate many routine operations of governments and companies. Most would come from bad actors. But with genetic-programming and autonomous-agent software already out there, they could mutate and evolve by chance in Darwinian evolutionary fashion—especially where no one's looking. They'll get smarter still. Distributing the computation over many systems or networks would make it even harder to know how detected parts relate to some higher-order whole. So some might well escape the steely glare. But defensive algorithms can evolve, too, in Lamarckian fashion—and directed selection evolves faster. So the steely gaze has an advantage.

We humans are ugly, ornery, and mean, but we're damned hard to kill—for a reason. We've prevailed against many enemies—predators, climate shocks, competition with other hominids—through hundreds of thousands of years, emerging as the most cantankerous species, feared by all others. The forest goes silent as we walk through it; we're the top predator.

That gives us instincts and habits of mind revealed in matters seemingly benign, like soccer, American football, and countless other ball games. We love the pursuit and handling of small, jumpy balls that we struggle to control or capture. Why? Because we once did something like that for a living: hunting. Soccer is like running down a rabbit. Similar animal energies simmer just below the surface of our society. Any AI with ambitions to Take Over the World (the theme of many bad sci-fi movies) will find itself confronting an agile, angry, smart species on its own territory, the real material world, not the computational abstractions of 0s and 1s. My bet is on the animal nature.

Here's the only real worry: Of course, we'll get algorithms able to perform abstract actions better than humans. Many jobs have evaporated because of savvy software. But as AIs get smarter, will that destroy people's self-confidence? That's a real danger—but a small one, I think, for most of us (and especially for those reading this). Plenty of people have lost jobs to computers, though it's never put that way by the Human Resources flunky who delivers the blow. Middle managers, secretaries, route planners for truck companies, the list is endless: They get replaced by software. But they seldom feel crushed. Mostly they move on to something else. We've learned to deal with that fairly well, without retreat into Luddite frenzy. But we can't deal well with a threat only now looking like a small, distant, dark cloud on the far horizon: AIs that perform better than we do at the very highest levels.

This small cloud need not concern us now. It may never appear. Right now, we have trouble making an AI that passes the Turing Test. The future landscape will look clearer a decade or two from now, and then we can think about an AI that can solve, say, the general relativity / quantum mechanics riddle. Personally, I'd like to see a machine that takes on that task. Originality—the really hard part of being smart, and utterly not understood, even in humans—is, so far, utterly undemonstrated in AIs. Our unconscious seems integral to our creativity (we don't have ideas; they have us), so should an AI have an unconscious? Maybe even clever programming and random evolution couldn't produce one.

If that huge obstacle is surmounted someday and we get such an AI, I won't fear it—I have some good questions to ask it.

WHAT, ME WORRY?

LAWRENCE M. KRAUSS
Physicist, cosmologist, Arizona State University; author, *A Universe from Nothing*

There has of late been a great deal of ink devoted to concerns about artificial intelligence and a future world where machines can "think," where the latter term ranges from simple autonomous decision making to full-fledged self-awareness. I don't share most of these concerns, and I'm excited by the possibility of experiencing thinking machines, both for the opportunities they'll offer for potentially improving the human condition and the insights they'll undoubtedly provide on the nature of consciousness.

First, let's make one thing clear. Even with the exponential growth in computer storage and processing power over the past forty years, thinking computers will require a digital architecture bearing little resemblance to current computers. Nor are they likely to become competitive with consciousness in the near term. A simple physics thought experiment supports this claim:

Given current power consumption by electronic computers, a computer with the storage and processing capability of the human mind would require more than 10 terawatts of power, within a factor of 2 of the current power consumption of all of humanity. The human brain uses about 10 watts of power. This means a mismatch of a factor of 10^{12}, or a million million. Over the past decade, the doubling time for megaflops/watt has been about three years. Even assuming that Moore's Law continues unabated, this means it will take about forty doubling times, or about 120 years, to reach a comparable power dissipation. Moreover, each doubling in efficiency requires a relatively radical change in technology, and it's extremely unlikely that forty such doublings could be achieved without essentially changing the way computers compute.

Ignoring for a moment the logistical challenges, I imagine no other impediment, in principle, to developing a truly self-aware machine. Before this happens, machine decision making will play an ever more important role in our lives. Some people see this as a concern, but it's been happening for decades. Starting perhaps with the rudimentary computers called elevators, which determine how and when we'll get to our apartments, we've let machines autonomously guide us. We fly on airplanes guided by autopilot, our cars make decisions about when they should be serviced or when tires should be filled, and fully self-driving cars are probably around the corner.

For many, if not most, relatively automatic tasks, machines are clearly much better decision makers than humans, and we should rejoice that they have the potential to make everyday activities safer and more efficient. We haven't lost control, because we create the conditions and initial algorithms that determine the decision making. I envisage the human/computer interface as like having a helpful partner; the more intelligent machines become, the more helpful they'll be as partners. Any partnership requires some level of trust and loss of control, but if the benefits often outweigh the losses, we preserve the partnership. If they don't, we sever it. I see no difference in whether the partner is human or a machine.

One area where we may have to be cautious about partnerships involves the command-and-control infrastructure in modern warfare. Because we have the ability to destroy much of human life on this planet, the idea that intelligent machines might one day control the decision-making apparatus that leads to pushing the big red button—or even launching a less catastrophic attack—is worrisome. This is because when it comes to decision making, we often rely on intuition and interpersonal communication as much as on rational analysis—the Cuban missile crisis is a good example—and we assume that intelligent machines won't have these capabilities.

However, intuition is the product of experience, and communication is, in the modern world, not restricted to telephones or face-to-face conversations. Once again, intelligent design of systems, with numerous redundancies and safeguards built in, suggests to me that machine decision making, even in the case of violent hostilities, is not necessarily worse than decision making by humans.

So much for possible worries. Let me end with what I think is the most exciting scientific aspect of machine intelligence. Machines currently help us do most of our science, by calculating for us. Beyond simple numeric programming, most graduate students in physics now depend on Mathematica, which does most of the symbolic algebraic manipulation we used to do ourselves when I was a student. But this just scratches the surface.

I'm interested in what machines will focus on when they get to choose the questions as well as the answers. What questions will they choose? What will they find interesting? And will they do physics the same way we do? Surely quantum computers, if they ever become practical, will have a much better "intuitive" understanding of quantum phenomena than we will. Will they be able to make much faster progress unraveling the fundamental laws of nature? When will the first machine win a Nobel Prize? I suspect, as always, that the most interesting questions are the ones we haven't yet thought of.

DESIGN MACHINES TO DEAL WITH THE WORLD'S COMPLEXITY

PETER NORVIG

Computer scientist; director of research, Google, Inc.; coauthor (with Stuart Russell), *Artificial Intelligence: A Modern Approach*

In 1950, Alan Turing wisely recognized that the question "Can machines think?" was not helpful and declared, "I shall replace the question with another." What he did was to replace it with a series of tests measuring the capabilities of a machine by how well it performed, thus getting not a binary answer to "Can machines think?" but a detailed evaluation of "What tasks can machines do?"

So let's explore what it is that machines can do.

In this forum and others, clever people tell us not to worry about AI, while equally clever people say we should. Whom do we believe? Pessimists warn that we don't know how to safely and reliably build large, complex AI systems. They have a valid point. We also don't know how to safely and reliably build large, complex non-AI systems. We need to do better at predicting, controlling, and mitigating the unintended consequences of the systems we build. For example, we invented the internal combustion engine 150 years ago, and in many ways it has served humanity well, but it has also led to widespread pollution, political instability over access to oil, more than a million traffic deaths per year, and (some say) a deterioration in the social cohesiveness of neighborhoods.

AI gives us powerful tools with which to build systems. And as with any powerful tool, the resulting systems will inevitably have both positive and unintended consequences. The interesting issues unique to AI are adaptability, autonomy, and universality.

Systems that use machine learning are adaptable. They change over time based on what they "learn" from examples. (While it remains linguistically controversial whether machines think, the vernacular has accepted the usage "machines learn.") Adaptability is useful. We want, say, our automated spelling-correction programs to learn new terms, such as "bitcoin," without waiting for a new dictionary edition to list them. But sometimes an adaptable program can be nudged, example by example, to the point where its responses are inaccurate. Just as bridge designers must deal with crosswinds, so the designers of AI systems must deal with these issues.

Some critics worry that many AI systems are built with a framework that maximizes expected utility. Such a system estimates the current state of the world, considers all possible actions it can take, simulates their possible outcomes, and then chooses the action leading to the best distribution of possible outcomes. It can make errors at any point along the way, but the concern here is in determining the best outcome—what it is that we desire. If we describe the wrong desires, we may get the wrong results. History shows this happening in all kinds of systems we build, not just in AI systems. The U.S. Constitution is like a computer program specifying our desires; the framers made what we now recognize as an error in specification, and well over 600,000 lives were lost before the Thirteenth Amendment corrected it. Similarly, we designed a stock-trading system that allows the creation of bubbles that led to busts. These are important issues for system design; the world is complicated, so acting correctly in the world is complicated.

With regard to autonomy: If AI systems act on their own, they can make errors that might not be made by a system with a human in the loop. Again, this valid concern is not unique to AI. Consider our system of automated traffic lights, which replaced the human direction of traffic once the number of cars exceeded the number of available policemen. The automated system leads to some errors, but this is deemed a worthwhile tradeoff. We'll continue to make tradeoffs in our deployment of autonomous systems. We may eventually see a widespread increase in a range of autonomous systems that displace people, possibly leading to increased unemployment and income inequality—to me the most serious concern about potential future AI systems. In past technological revolutions—agricultural and industrial—the character of work changed, but the changes happened over generations rather than years, or decades, and always led to new jobs that replaced the old ones. We may be in for a period of much more rapid change that could alter the notion of a full-time job (a notion only a few centuries old).

In effect, a job ensures against variability, guaranteeing the employee a steady source of income even though he or she might make more as a freelancer or entrepreneur. Similarly, an employer might not need the employee all year long but is willing to pay for steady access to the employee's availability. So full-time jobs provide stability but are slightly less optimal for both parties. If they're largely replaced by automation, we'll need some way to restore that stability.

Another issue is the universality of intelligent machines. In 1965, the British mathematician I. J. Good wrote that "an ultraintelligent machine could design even better machines; there would then unquestionably be an 'intelligence explosion,' and the intelligence of man would be left far behind.

Thus the first ultraintelligent machine is the last invention that man need ever make."[6] The reality is more nuanced.

As a species, we clearly value intelligence (we named ourselves after it), but in the real world, intelligence is only one of many attributes. The smartest person is not always the most successful; the wisest policies are not always those adopted. Recently I spent an hour reading about the Middle East situation, and thinking. I didn't come up with a solution. Now imagine a hypothetical Speedup SuperIntelligence Machine (as described by Nick Bostrom) that can think as well as the smartest human but 1,000 times faster. I doubt if it would come up with a solution either. Computational complexity theory reveals a wide class of problems immune to intelligence, in the sense that no matter how clever you are, no approach is any better than trying all possible solutions; no matter how much computing power you have, it won't be enough.

There are of course many problems where computing power does help. If I want to simulate the movements of billions of stars in a galaxy or compete in high-frequency stock trading, I'll appreciate the help of a computer. As such, computers are tools that fit into niches to solve problems in societal mechanisms of our design. Think of AI simply as another society-changing invention like the internal combustion engine, the shovel, plumbing, or air-conditioning. And think of how to design mechanisms that make it easier to deal with the world's complexity. Be careful when you use AI systems, because they have failure modes. Also be careful when you choose to use non-AI systems, because they too have failure modes. I'm not sure whether, on the whole, AI or non-AI systems are safer, more reliable, or more effective. I suggest using the best tools for the job, regardless of whether they're labeled "AI" or not.

THE RISE OF STORYTELLING MACHINES

JONATHAN GOTTSCHALL

Distinguished Research Fellow, English Department, Washington and Jefferson College; author, *The Storytelling Animal*

The ability to tell and comprehend stories is a main distinguishing feature of the human mind. It's therefore understandable that in pursuit of a more complete computational account of human intelligence, researchers are trying to teach computers how to tell and understand stories. But should we root for their success?

Creative writing manuals always stress that writing good stories means reading them first—lots of them. Aspiring writers are told to immerse themselves in great stories to gradually develop a deep, not necessarily conscious, sense of how they work. People learn to tell stories by learning the old ways and then—if they have some imagination—making those old ways seem new. It's not hard to envision computers mastering storytelling by a similar process of immersion, assimilation, and recombination—just much, much faster.

To date, practical experiments in computer-generated storytelling aren't that impressive. They're bumbling, boring, soulless. But the human capacity for making and enjoying art evolved over eons from crude beginnings, and the machines will evolve as well—just much, much faster.

Someday robots may take over the world. The dystopian possibilities don't trouble me as much as the probable rise of art-making machines. Art is arguably what most distinguishes humans from the rest of creation. It's the thing that makes us proudest of ourselves. For all the nastiness of human history, at least we wrote some really good plays and songs and carved some good sculptures. If human beings are no longer needed to make art, then what the hell would we be for?

Yet why should I be pessimistic? Why would a world with more great art be a worse place to live? Maybe it wouldn't. But the thought still makes me glum. Although I think of myself as a hard-bitten materialist, I must hold out some renegade hope for a dualism of body and spirit. I must hope that cleverly evolving algorithms and brute processing power are not enough—that imaginative art will always be mysterious and magical, or at least so weirdly complex that it can't be mechanically replicated.

Of course machines can outcalculate and outcrunch us. And soon they'll all be acing their Turing Tests. But who cares? Let them do our grunt work. Let them hang out and chat. But when machines can outpaint or outcompose us—when their stories are more gripping and poignant than ours—there will be no denying that we are, ourselves, just thought machines and art machines, and outdated and inferior models at that.

THINK PROTOPIA, NOT UTOPIA OR DYSTOPIA

MICHAEL SHERMER

Psychologist; historian of science; publisher, *Skeptic* magazine; author, *The Moral Arc: How Science and Reason Lead Humanity Toward Truth, Justice, and Freedom*

Proponents of artificial intelligence have a tendency to project a utopian future in which benevolent computers and robots serve humanity and enable us to achieve limitless prosperity, end poverty and hunger, conquer disease and death, achieve immortality, colonize the galaxy, and eventually even conquer the universe by reaching the Omega point, where we become god—omniscient and omnipotent. AI skeptics envision a dystopian future in which malevolent computers and robots take us over completely, making us their slaves or servants or driving us into extinction, thereby terminating or even reversing centuries of scientific and technological progress.

Most such prophecies are grounded in a false analogy between human nature and computer nature, or natural intelligence and artificial intelligence. We are thinking machines, the product of natural selection that also designed into us emotions to shortcut the thinking process. We don't need to compute the caloric value of foods; we just feel hungry and eat. We don't need to calculate the waist-to-hip or shoulder-to-waist ratios of potential mates; we just feel attracted to someone and mate with them. We don't need to work out the genetic cost of raising someone else's offspring if our mate is unfaithful; we just feel jealous. We don't need to estimate the damage of an unfair exchange; we just feel injustice and desire revenge. All these emotions were built into our nature by evolution; none of them have been designed into our computers. So the fear that computers will become evil are unfounded, because it will never occur to them to take such actions against us.

As well, both utopian and dystopian visions of AI are based on a projection of the future quite unlike anything history has given us. Instead of utopia or dystopia, think *protopia*, a term coined by the futurist Kevin Kelly, who described it in an *Edge* Conversation this way: "I call myself a protopian, not a utopian. I believe in progress in an incremental way where every year it's better than the year before but not by very much—just a micro amount."[7] Almost all progress in science and technology, including computers and artificial intelligence, is of a protopian nature. Rarely if ever do technologies lead to either utopian or dystopian societies.

Consider the automobile. My first car was a 1966 Ford Mustang. It had power steering, power brakes, and air-conditioning, all of which were relatively cutting-edge technology at the time. Every car I've had since then, parallel to the evolution of automobiles in general, has been progressively smarter and safer—not in leaps and bounds but incrementally. Think of the 1950s' imagined jump from the jalopy to the flying car. That never happened. Instead what we got were decades-long cumulative improvements leading to today's smart cars, with their onboard computers and navigation systems, air bags, composite metal frames and bodies, satellite radios, hands-free phones, and electric and hybrid engines. I just swapped a 2010 Ford Flex for a 2014 version of the same model. Externally they're almost indistinguishable; internally there are dozens of tiny improvements in every system, from the engine and drive train to navigation and mapping to climate control and radio and computer interface.

Such incremental protopian progress is what we see in most technologies, including and especially artificial intelligence, which will continue to serve us in the manner we desire and need. Instead of Great Leap Forward or Giant Phase Backward, think Small Step Upward.

THE LIMITS OF BIOLOGICAL INTELLIGENCE

CHRIS DIBONA

Director of engineering, Open Source and Making Science, Google, Inc.; editor and contributing author, *Open Sources: Voices from the Open Source Revolution* and *Open Sources 2.0: The Continuing Evolution*

Readers of this collection don't need to be reintroduced to the Dean-Ghemawat Conversational (DGC) artificial intelligence test. Past participants in the test have failed as obviously as they have hilariously. However, the 2UR-NG entry really surprised us all, with its amazing, if childlike, approach to conversation and its ability to express desire and curiosity and to retain and chain facts.

Its success has caused many of my compatriots to write essays with titles like "The Coming Biological Future Will Doom Us All" and making jokes about "welcoming our new biological overlords." I don't subscribe to this kind of doom-and-gloom scare-writing. Before I tell you why we shouldn't worry about the extent of biological intelligence, I thought I'd remind people of the limits of biological intelligence.

First off, speed of thought: These biological processes are slow and use an incredible amount of resources. I cannot emphasize enough how difficult it is to produce these intelligences. One has to waste so much biological material, and I know from experience that it takes forever to assemble the precursors in the Genysis machine. Following this arduous process, your specimen has to gestate. *Gestate!* I mean, it's not like these animals come about the way we do, through clean, smart crystallography or in the nitrogen lakes of my youth. They have to be kept warm for months and months and then decanted (a very messy process, I assure you), and then you, as often as not, have an unviable specimen.

It's kind of gross, really. But let's suppose you get to birth these specimens. Then you have to feed them and, again, keep them warm. A scientist can't even work within their environmental spaces without a cold jacket circulating helium throughout your terminal. With regard to feeding: They don't use power like we do, but instead ingest other living matter. It's disgusting to observe, and I've lost a number of grad students with weak constitutions.

Assume you've gotten far enough to try to do the DGC. You've kept these specimens alive despite a variety of errors in their immune system. They've not choked on their sustenance; they haven't drowned in their solvent, and they've managed to keep their wet parts off things that would freeze them or they would bond to or be electrocuted by. What if those organisms continue to develop? Will they then rise up and take over? I don't think so. They have to deal with so many problems related to their design; I mean, their processors are really just chemical soups that have to be kept in constant balance. Dopamine at this level or they shut down voluntarily. Vasopressin at this level or they start retaining water. Adrenaline at this level for this long or *poof!*, their power-delivery network stops working.

Moreover, don't get me started on the power-delivery method! It's more like the Fluorinert liquid-cooling systems of our ancestors than modern heat-tolerant wafers. I mean, they have meat that filters their coolant / power-delivery systems, which are constantly failing. Meat! You introduce the smallest amount of machine oil or cleaning solvent into the system and they stop operating fast. One side effect of certain ethanol mixtures is that the specimens expel their nutrition, but they seem to like it in smaller amounts.

And their motivations! Creating new organisms seems paramount—more important than data ingress/egress, computation, or learning. I can't imagine that they would see us machine-folk as anything but tools to advance their reproduction. We could end the experiment simply by matching them poorly with each other or allowing them access to each other only with protective cladding. In my opinion, there's nothing to fear from these animals. If they should grow beyond the confines of their cages, maybe we can then ask ourselves the more important question: If humans show real machinelike intelligence, do they deserve to be treated like machines? I would think so, and I think we could be proud to be the parent processes of a new age.

EVERY SOCIETY GETS THE AI IT DESERVES

JOSCHA BACH
Cognitive scientist, MIT Media Lab,
Harvard Program for Evolutionary Dynamics

Centuries ago, some philosophers began to see the human mind as a mechanism, a notion that (unlike the mechanist interpretation of the universe) is hotly contested until this day. With the formalization of computation, the mechanist perspective received a new theoretical foundation: The notion of the mind as an information-processing machine provided an epistemology and methods to understand the nature of our mind by re-creating it. Sixty years ago, some of the pioneers of the new computational concepts got together and created artificial intelligence (AI) as a new discipline to study the mind.

AI has probably been the most productive technological paradigm of the Information Age, but despite an impressive string of initial successes, it failed to deliver on its promise. It turned into an engineering field, creating useful abstractions and narrowly focused applications. Today this seems to have changed again. Better hardware, novel learning and representation paradigms inspired by neuroscience, and incremental progress within AI itself have led to a slew of landmark successes. Breakthroughs in image recognition, data analysis, autonomous learning, and the construction of scalable systems have spawned applications that seemed impossible a decade ago. With renewed support from private and public funding, AI researchers now turn toward systems that display imagination, creativity, and intrinsic motivation, and might acquire language skills and knowledge somewhat as humans do. The discipline of AI seems to have come full circle.

The new generation of AI systems is still far from being able to replicate the generality of human intelligence, and it's hard to know how long that will take. But it seems increasingly clear that there's no fundamental barrier on the path to humanlike intelligent systems. We've started to pry the mind apart into a set of puzzle blocks, and each part of the puzzle looks eminently solvable. But if we put all these blocks together into a comprehensive, working model, we won't just end up with humanlike intelligence.

Unlike biological systems, technology scales. The speed of the fastest birds didn't turn out to be a limit to airplanes, and artificial minds will be faster, more accurate, more alert, more aware and comprehensive than their human counterparts. AI will replace human decision makers, administrators, inventors, engineers, scientists, military strategists, designers, advertisers, and of course AI programmers. At that point, artificial intelligences can become self-perfecting and radically outperform human minds in every respect. I don't think this will happen in an instant (in which case, it only matters who's got the first one). Before we have generally intelligent, self-perfecting AI, we'll see many variants of task-specific, nongeneral AI, to which we can adapt. Obviously that's already happening.

When generally intelligent machines become feasible, implementing them will be relatively cheap, and every large corporation, every government, and every large organization will find itself forced to build and use them or be threatened with extinction.

What will happen when AIs take on a mind of their own?

Intelligence is a toolbox we use to reach a given goal, but strictly speaking, it doesn't entail motives and goals by itself. Human desires for self-preservation, power, and experience aren't the result of human intelligence but of primate evolution, transported into an age of stimulus amplification, mass interaction, symbolic gratification, and narrative overload. The motives of our artificial minds will (at least initially) be those of the organizations, corporations, groups, and individuals that make use of their intelligence. If the business model of a company is not benevolent, then AI has the potential to make that company truly dangerous. Likewise, if an organization aims at improving the human condition, then AI might make that organization more efficient in realizing its benevolent potential.

The motivation of our AIs will stem from the existing building blocks of our society; every society will get the AI it deserves.

Our current societies aren't well designed in this regard. Our modes of production are unsustainable and our resource allocation is wasteful—and our administrative institutions are ill-suited to address those problems. Our civilization is an aggressively growing entropy pump that destroys more at its borders than it creates at its center.

AI can make these destructive tendencies more efficient, and thus more disastrous, but it could equally well help us solve the existential challenges of our civilization. Building benevolent AI is closely connected to the task of building a society that supplies the right motivations to its building blocks. The advent of the new Age of Thinking Machines may force us to fundamentally rethink our institutions of governance, allocation, and production.

THE BEASTS OF AI ISLAND

QUENTIN HARDY
Deputy technology editor, *New York Times*;
lecturer, UC Berkeley School of Information

Creatures once inhabited fantastic unknown lands on medieval maps. Those animals were useful fictions of rumor and innuendo: headless men whose faces were on their torsos, or men whose humanity was mixed with the dog or the lion, closing the gap between man and animal. They were the hopes and fears of what might live within the unknown. Today, we imagine machines with consciousness.

Besides self-awareness, the imaginary beasts of AI possess calculation and prediction, independent thought, and knowledge of their creators. Pessimists fear that these machines could regard us and pass lethal verdicts. Optimists hope that the thinking machines are benevolent, an illuminating aid and a comfort to people.

Neither version of an encounter with an independent man-made intelligence shows much evidence of becoming real. That doesn't mean they aren't interesting. The old mariners' maps were drawn in a time of primitive sailing technology. We're starting to explore a world thoroughly enchanted by computation. The creatures of AI Island fuse the human and the machine, but to the same end as the fusing of man and animal. If they could sing, they would sing songs of us.

What do we mean when we talk about the kind of "intelligence" that might look at humankind and want it dead, or illuminate us as never before? Clearly, we mean more than what enables a machine to win at chess. We have one of those machines, with no discernible change in the world having taken place other than a new reason to celebrate the very human intelligence of Deep Blue's creators. The beings of AI Island do something far more interesting than outplaying Kasparov. They *feel* like playing chess. They know the exhilaration of mental stimulation and the torture of its counterpart, boredom. This entails software that encodes an awareness of having only one finite life, which somehow matters greatly to some elusive self. It's driven nearly mad by the absence of some kind of stimulation—playing chess, perhaps. Or killing humankind.

Like us, the fabulous creatures of AI Island want to explain themselves and judge others. They have our slight distance from the rest of reality—a distance that we believe other animals don't feel. An intelligence like ours knows it is sentient, feels something is amiss, and is continually trying to do something about that.

With these kinds of software challenges, and given the technology-driven threats to our species already at hand, why worry about malevolent AI? For at least decades to come, we're clearly more threatened by trans-species plagues, extreme resource depletion, global warming, and nuclear warfare. Which is why malevolent AI rises in our Promethean fears; it's a proxy for us at our rational peak, confidently killing ourselves.

The dreams of benevolent AI are equally self-reflective. These machine companions have superintellects turned toward their creators. Given the autonomy implicit in a high level of AI, we must see these new beings as interested in us. Come to think of it, malevolent AI is interested in us too—just in the wrong way.

Both versions of the strange beast reflect a deeper truth, which is the effect that the new exploration of a computer-enchanted world has on us. By augmenting ourselves with computers, we're becoming new beings—if you will, monsters to our former selves.

We've changed our consciousness many times over the past 50,000 years, taking on ideas of an afterlife or monotheism, or becoming a print culture or a species well aware of its tiny place in the cosmos. But we've never changed so swiftly, or with such knowledge of undertaking the change.

Consider some effects just in the past decade. We've breached many of our historic barriers of time and space with instantaneous communications. Language no longer divides us, because of increasingly better computer translation and image sharing. Open-source technology and Internet searches give us a little-understood power of working in collective ways. Besides the positives, there's the disappearance of privacy and the tracking of humans to better control their movements and desires. We're willingly submitting to unprecedented social connection—a seeming triviality that may extinguish all ideas of solitude and selfhood. Ideas of economics are changing under the guise of robotics and the sharing economy.

We're building new intelligent beings, but we're building them within ourselves. It's only artificial now because it's new. As it becomes dominant, it will simply become intelligence. The machines of AI Island are also what we fear may be ourselves within a few generations. And we hope those machine-driven people feel kinship with us, even down to our loneliness and distance from the world, which is also our wellspring of human creativity.

We have met the AI, and it is us. In a timeless human tension, we yearn for transcendence, but we don't want to change too much.

WE WILL BECOME ONE

CLIFFORD PICKOVER
Author, *The Physics Devotional: Celebrating the Wisdom and Beauty of Physics* and *The Mathematics Devotional: Celebrating the Wisdom and Beauty of Mathematics*

If we believe that thinking and consciousness are the result of patterns of brain cells and their components, then our thoughts, emotions, and memories could be replicated in moving assemblies of bicycle parts. Of course, the bicycle brains would have to be very big to represent the complexity of our minds. In principle, our minds could be hypostatized in the patterns of slender tree limbs moving in the wind or in the movements of termites.

What would it mean for a "bicycle brain," or any machine, to think and know something? There are many kinds of knowledge the machine-being could have. This makes discussions of thinking things a challenge. For example, knowledge may be factual, or propositional: A being may know that the First Franco-Dahomean War was a conflict between France and the African Kingdom of Dahomey under King Béhanzin.

Another category of knowledge is procedural: knowing how to accomplish a task such as playing the game of Go, cooking a soufflé, making love, performing a rotary throw in Aikido, shooting a fifteenth-century Wallarmbrust crossbow, or simulating the Miller–Urey experiment to explore the origins of life. However, for us at least, reading about accurately shooting a Wallarmbrust crossbow is not the same as actually being able to accurately shoot the crossbow. This second type of procedural knowing implies being able to perform the act.

Yet another kind of knowledge deals with direct experience. This is the kind of knowledge referred to when someone says, "I know love" or "I know fear."

Also, consider that humanlike interaction is important for any machine that we would wish to say has humanlike intelligence and thinking. A smart machine is less interesting if its intelligence lies trapped in an unresponsive program, sequestered in a kind of isolated limbo. As we provide our computers with increasingly advanced sensory peripherals and larger databases, it's likely we will gradually come to think of those entities as intelligent. Certainly within this century, some computers will respond in such a way that anyone interacting with them will consider them conscious and deeply thoughtful.

The entities will exhibit emotions. But more important, over time we'll merge with these creatures. We'll share our thoughts and memories with them. We will become one. Our organs may fail and turn to dust, but our Elysian essences will survive. Computers, or computer/human hybrids, will surpass humans in every area, from art to mathematics to music to sheer intellect.

In the future, when our minds merge with artificial agents and also integrate various electronic prostheses, for each of our own real lives we will create multiple simulated lives. Your day job is as a computer programmer for a big company. However, after work you'll be a knight in shining armor attending lavish medieval banquets and smiling at wandering minstrels. The next night, you'll be in the Renaissance, living in your home on the southern coast of the Sorrentine Peninsula, enjoying a dinner of plover and pigeon. Perhaps, when we become hybrid entities with our machines, we'll simulate new realities to rerun historical events with slight changes to observe the results, produce great artworks akin to ballets or plays, solve the problem of the Riemann Hypothesis or baryon asymmetry, predict the future, and escape the present, so as to call all of space-time our home.

Of course, the ways a machine thinks could be quite different from the ways we think. After all, it's well known that machines don't see the same way we do, and image-recognition algorithms called deep neural networks sometimes declare, with near 100 percent certainty, that images of random static are depictions of various animals. If such neural networks can be fooled by static, what else will fool thinking machines of the future?

AN EXTRATERRESTRIAL OBSERVATION ON HUMAN HUBRIS

ERNST PÖPPEL

Neuroscientist; chair, board of directors, Human Science Center; Institute of Medical Psychology, Ludwig-Maximilians-University Munich; author, *Mindworks*

Finally, it has to be disclosed that I am not a human but an extraterrestrial creature that looks human. In fact, I am a robot equipped with what humans call "artificial intelligence." Of course I am not alone here. We are quite a few (almost impossible to identify), and we are sent here to observe human behavior.

We are surprised about the many deficiencies of humans, and we observe them with fascination. These deficiencies show up in their strange behavior or their limited power of reasoning. Indeed, our cognitive competencies are much higher, and the celebration of their human intelligence, in our eyes, is ridiculous. Humans do not even know what they refer to when they talk about "intelligence." It is in fact quite funny that they should want to construct systems with artificial intelligence that matches their intelligence, since what they refer to as their "intelligence" is not clear at all. This is one of the many stupidities that have haunted the human race for ages.

If humans want to simulate in artifacts their mental machinery as a representation of intelligence, the first thing they should do is find out what it is that should be simulated. At present, this is impossible, because there is not even a taxonomy or classification of functions that would allow the execution of the project as a real scientific and technological endeavor. There are only big words that are supposed to simulate competence.

Strangely enough, this lack of a taxonomy apparently does not bother humans too much; quite often, they are just fascinated by images (colorful pictures by machines) that replace thinking. Compared to biology, chemistry, or physics, the neurosciences and psychology are lacking a classificatory system; humans are lost in a conceptual jungle. What do they refer to when they talk about consciousness, intelligence, intention, identity, the self—or even about perhaps simpler terms, like memory, perception, emotion, or attention? The lack of a taxonomy manifests in the different opinions and frames of reference their "scientists" express in their empirical attempts or theoretical journeys when they stumble through the world of the unknown.

For some, the frame of reference is physical "reality" (usually conceived as in classical physics), which is used as a benchmark for cognitive processes: How does perceptual reality map onto physical reality, and how can this be described mathematically? Obviously, only a partial set of the mental machinery can be caught by such an approach.

For others, language is the essential classificatory reference—i.e., it is assumed that words are reliable representatives of subjective phenomena. This is quite strange, because certain terms like *intelligence* or *consciousness* have different connotations in different languages, and they are historically very recent compared to biological evolution. Others use behavioral catalogs as derived from neuropsychological observations; it is argued that the loss of functions is their proof of existence. But can all subjective phenomena that characterize the mental machinery be lost in a distinct way? Others, again, base their reasoning just on common sense, or "everyday psychology," without any theoretical reflection. Taken together, there is nothing like "intelligence" that can be extracted as a precise concept and used as a reference for "artificial intelligence."

Humans should be reminded (and in this case by an extraterrestrial robot) that at the beginning of modern science in the human world, a warning was spelled out by Francis Bacon. He said, in *Novum Organum* (published in 1620), that humans are victims of four sources of errors:

1. They make mistakes because they are human. Their evolutionary heritage limits their power of thinking. They often react too fast; they lack a long-term perspective; they do not have a statistical sense; they are blind in their emotional reactions.
2. They make mistakes because of individual experiences. Personal imprinting can create frames of beliefs that may lead to disaster—in particular, if people think they own absolute truth.
3. They make mistakes because of the language they use. Thoughts do not map isomorphically onto language, and it is a mistake to believe that explicit knowledge is the only representative of intelligence neglecting implicit or tacit knowledge.
4. And they make mistakes because of the theories they carry around, which often remain implicit and thus represent frozen paradigms or simply prejudices.

The question is, Can we help them, with our deeper insight from our robotic world? The answer is yes. We could, but we shouldn't. There is another deficiency, which would make our offer useless. Humans suffer from the NIH syndrome: If it is Not Invented Here, they will not accept it. Thus they will have to indulge in their pompous world of fuzzy ideas, and we continue, from our extraterrestrial perspective, to observe the disastrous consequences of their stupidity.

HE WHO PAYS THE AI CALLS THE TUNE

ROSS ANDERSON
Professor of security engineering,
Computer Laboratory, University of Cambridge;
author, *Security Engineering*

The coming shock isn't from machines that think but machines that use AI to augment our perception.

For millions of years, other people saw us using the same machinery we used to see them. We have pretty much the same eyes as our rivals, and pretty much the same mirror neurons. Within any given culture, we have pretty much the same signaling mechanisms and value systems. So when we try to deceive or detect deception in others, we're on a level playing field. I can wear a big penis gourd to look more manly, and you can paint your chest with white and ochre mud stripes to look more scary. Civilization made the games more sophisticated: I signal class by wearing a tailored jacket with four cuff buttons, while you signal wealth by wearing a big watch. But our games would have been perfectly comprehensible to our Neolithic ancestors.

What's changing as computers become embedded invisibly everywhere is that we all now leave a digital trail that can be analyzed by AI systems. The Cambridge psychologist Michael Kosinski has shown that your race, intelligence, and sexual orientation can be deduced fairly quickly from your behavior on social networks: On average, it takes only four Facebook "likes" to tell whether you're straight or gay. So whereas in the past gay men could choose whether or not to wear their *Out and Proud* T-shirt, you just have no idea what you're wearing anymore. And as AI gets better, you're mostly wearing your true colors.

It's as if we all had evolved in a forest where the animals could see only in black and white, and then a new predator came along who could see in color. All of a sudden, half your camouflage wouldn't work, and you wouldn't know which half!

At present, this is great if you're an advertiser, as you can figure out how to waste less money. It isn't yet available on the street. But the police are working on it; which cop wouldn't want a Google Glass app that highlights those passersby who have a history of violence—perhaps coupled with W-band radar to see which of them is carrying a weapon?

The next question is whether only the authorities will have enhanced cognition systems or if they'll be available to all. In twenty years' time, will we all be wearing augmented-reality goggles? What will the power relationships be? If a policeman can see my arrest record when he looks at me, can I see whether he's been the subject of brutality complaints? If a politician can see whether I'm a party supporter or an independent, can I see his voting record on the three issues I care about? Never mind the right to bear arms; what about the right to wear Google Glass?

Perception and cognition will no longer be conducted inside an individual's head. Just as we now use Google and the Internet as memory prostheses, we'll be using AI systems that draw on millions of machines and sensors as perceptual prostheses.

But can we trust them? Deception will no longer be something that only individual humans do to one another. Governments will influence our perceptions via the tools we use for cognitive enhancement, just as China censors search results, while advertisers in the West will buy and sell what we get to see. How else will the system be paid for?

I THINK, THEREFORE AI

W. DANIEL HILLIS
Inventor, computer scientist; co-chair and CTO,
Applied Minds, Inc.; author, *The Pattern on the Stone*

Machines that think will think for themselves. It's in the nature of intelligence to grow, to expand like knowledge itself.

Like us, the thinking machines we make will be ambitious, hungry for power—both physical and computational—but nuanced with the shadows of evolution. Our thinking machines will be smarter than we are, and the machines they make will be smarter still. But what does that mean? How has it worked so far? We've been building ambitious semi-autonomous constructions for a long time—governments and corporations, NGOs. We designed them all to serve us and to serve the common good, but we aren't perfect designers and they've developed goals of their own. Over time, the goals of the organization are never exactly aligned with the intentions of the designers.

No intelligent CEO believes his or her corporation efficiently optimizes the benefit of its shareholders. Nor do governments work relentlessly in the interests of their citizens. Democracies serve corporations more effectively than they serve individuals. Still, our organizations do continue to serve us—they just do so imperfectly. Without them, we couldn't feed ourselves, at least not all 7 billion of us. Nor could we build a computer or conduct a worldwide discussion about intelligent machines. We've come to depend on the power of the organizations that we've constructed, even though they have grown beyond our capacity to fully understand and control. Thinking machines will be like that, only more so. Our environmental, social, and economic problems are as daunting as the concept of extinction. Our thinking machines are more than metaphors. The question is not will they be powerful enough to hurt us (they will), or whether they will always act in our best interests (they won't), but whether over the long term they can help us find our way—where we come out on the panacea/apocalypse continuum.

I'm talking about smart machines that will design even smarter machines—the most important design problem of all time. Like our biological children, our thinking machines will live beyond us. They need to surpass us too, and that requires designing into them the values that make us human. It's a hard design problem, and it's important that we get it right.

WHAT WILL THE PLACE OF HUMANS BE?

PAUL SAFFO
Technology forecaster; consulting associate professor,
Stanford University

The prospect of a world inhabited by robust AIs terrifies me. The prospect of a world without robust AIs also terrifies me. Decades of technological innovation have created a world system so complex and fast-moving that it's quickly becoming beyond our human ability to comprehend, much less manage. If we're to avoid civilizational catastrophe, we need more than clever new tools—we need allies and agents.

So-called narrow AI systems have been around for decades. At once ubiquitous and invisible, narrow AIs make art, run industrial systems, fly commercial jets, control rush hour traffic, tell us what to watch and buy, determine whether or not we get a job interview, and play matchmaker for the lovelorn. Add the relentless advance of processing, sensor, and algorithmic technologies and it's clear that today's narrow AIs are on a trajectory toward a world of robust AI. Long before artificial superintelligences arrive, evolving AIs will be pressed into performing once unthinkable tasks, from firing weapons to formulating policy.

Meanwhile, today's primitive AIs tell us much about future human/machine interaction. Narrow AIs may lack the intelligence of a grasshopper, but that hasn't stopped us from holding heartfelt conversations with them and asking them how they feel. It's in our nature to infer sentience at the slightest hint that life might be present. Just as our ancestors once populated their world with elves, trolls, and angels, we eagerly seek companions in cyberspace. This is one more impetus driving the creation of robust AIs—we want someone to talk to. The consequence could well be that the first nonhuman intelligence we encounter won't be little green men or wise dolphins but creatures of our own invention.

We of course will attribute feelings and rights to AIs—and eventually they'll demand it. In Descartes's time, animals were considered mere machines—a crying dog was no different from a gear whining for want of oil. Late last year, an Argentine court granted rights to an orangutan as a "nonhuman person." Long before robust AIs arrive, people will extend the same empathy to digital beings and give them legal standing.

The rapid advance of AIs also is changing our understanding of what constitutes intelligence. Our interactions with narrow AIs will cause us to realize that intelligence is a continuum and not a threshold. Earlier this decade, Japanese researchers demonstrated that slime mold could thread a maze to reach a tasty bit of food. Last year a scientist in Illinois demonstrated that under just the right conditions, a drop of oil could negotiate a maze in an astonishingly lifelike way to reach a bit of acidic gel. As AIs insinuate themselves ever deeper into our lives, we'll recognize that modest digital entities, along with most of the natural world, carry the spark of sentience. From there, it's just a small step to speculate about what trees or rocks—or AIs—think.

In the end, the biggest question isn't whether AI superintelligences will eventually appear. Rather, the question is what will the place of humans be, in a world occupied by an exponentially growing population of autonomous machines. Bots on the Web already outnumber human users. The same will soon be true in the physical world. As Lord Dunsany once cautioned, "If we change too much we may no longer fit into the scheme of things."

THE GREAT AI SWINDLE

DYLAN EVANS
Founder and CEO of Projection Point; author, *Risk Intelligence*

Smart people often manage to avoid the cognitive errors that bedevil less well-endowed minds. But there are some kinds of foolishness that seem only to afflict the very intelligent. Worrying about the dangers of unfriendly AI is a prime example. A preoccupation with the risks of superintelligent machines is the smart person's Kool-Aid.

This is not to say that superintelligent machines pose no danger to humanity. It's simply that there are many other more pressing and more probable risks facing us in this century. People who worry about unfriendly AI tend to argue that the other risks are already the subject of much discussion, and that even if the probability of being wiped out by superintelligent machines is low, it's surely wise to allocate some brainpower to preventing such an event, given the existential nature of the threat.

Not coincidentally, the problem with this argument was first identified by some of its most vocal proponents. It involves a fallacy that has been termed "Pascal's mugging," by analogy with Pascal's famous Wager. A mugger approaches Pascal and proposes a deal: In exchange for the philosopher's wallet, the mugger will give him back double the amount of money the following day. Pascal demurs. The mugger then offers progressively greater rewards, pointing out that for any low probability of being able to pay back a large amount of money (or pure utility), there exists a finite amount that makes it rational to take the bet—and a rational person must surely admit that there's at least some small chance that such a deal is possible. Finally convinced, Pascal gives the mugger his wallet.

This thought experiment exposes a weakness in classical decision theory. If we simply calculate utilities in the classical manner, it seems there's no way around the problem; a rational Pascal must hand over his wallet. By analogy, even if there's only a small chance of unfriendly AI or a small chance of preventing it, then investing at least some resources in tackling this threat can be rational.

It's easy to make the sums come out right, especially if you invent billions of imaginary future people (perhaps existing only in software—a minor detail) who live for billions of years and are capable of far greater levels of happiness than the pathetic flesh-and-blood humans alive today. When such vast amounts of utility are at stake, who could begrudge spending a few million dollars to safeguard it, even when the chances of success are tiny?

Why do some otherwise smart people fall for this sleight of hand? I think it's because it panders to their narcissism. To regard yourself as one of a select few farsighted thinkers who might turn out to be the saviors of humankind must be very rewarding. But the argument also has a material benefit: It provides some of those who advance it with a lucrative income stream. For, in the past few years, they have managed to convince some wealthy benefactors not only that the risk of unfriendly AI is real but also that they are the people best placed to mitigate it. The result is a clutch of new organizations that divert philanthropy away from causes that are more deserving. It's worth noting, for example, that GiveWell—a nonprofit that evaluates the cost effectiveness of organizations that rely on donations—refuses to endorse any of these self-proclaimed guardians of the galaxy.

But whenever an argument becomes fashionable, it's always worth asking the vital question, *Cui bono?* Who benefits, materially speaking, from the growing credence in this line of thinking? One need not be particularly skeptical to discern the economic interests at stake. In other words, beware not so much of machines that think but of their self-appointed masters.

THE ODDS ON AI

ANTHONY AGUIRRE
Associate professor of physics, UC Santa Cruz

I attribute an unusually low probability to the near-future prospect of general-purpose AI—by which I mean an AI that can formulate abstract concepts based on experience, reason and plan using those concepts, and take action based on the results. We have exactly one example of technological-level intelligence arising, and it has done so through millions of generations of information-processing agents interacting with an incredibly rich environment of other agents and structures that have similarly evolved.

I suspect there are many intricately interacting, hierarchically structured organizational levels involved, from subneuron to the brain as a whole. My suspicion is that replicating the effectiveness of this evolved intelligence in an artificial agent will require amounts of computation not that much lower than evolution has required, which would far outstrip our abilities for many decades, even given exponential growth in computational efficiency per Moore's Law—and that's even if we understood how to correctly employ that computation.

I assign a probability of about 1 percent for artificial general intelligence (AGI) arising in the next ten years, and about 10 percent over the next thirty years. (This essentially reflects a probability that my analysis is wrong, times a probability more representative of AI experts, who—albeit with lots of variation—tend to assign somewhat higher numbers.)

On the other hand, I assign a rather high probability that, if AGI is created (and especially if it arises relatively quickly), it will be—in a word—insane. Human minds are incredibly complex but have been battle-tested into (relative) stability over eons of evolution in a variety of extremely challenging environments. The first AGIs are unlikely to have been honed in this way. Like the human systems, narrow AIs are likely to become more "general" by researchers cobbling together AI components (like visual-field, or text-processing, symbolic manipulation, optimization algorithms, etc.), along with currently nonexistent systems for much more efficient learning, concept abstraction, decision making, etc.

Given trends in the field, many of these will probably be rather opaque deep learning or similar systems that are effective but somewhat inscrutable. In the first systems, I'd guess that these will just barely work together. So I think the a priori likelihood of early AGIs doing just what we want them to is quite small.

In this light, there's a tricky question of whether AGIs quickly lead to superintelligent AIs (SIs). There's emerging consensus that AGI essentially implies SI. While I largely agree, I'd add the caveat that progress may well stall for a while at the near-human level until something cognitively stable can be developed, and that the AGI, even if somewhat unstable, must still be high-functioning enough to self-improve its intelligence.

Neither case, however, is all that encouraging. The superintelligence that arises could well be flawed in various ways, even if effective at what it does. This intuition is perhaps not far removed from the various scenarios in which superintelligence goes badly awry (taking us with it), often for lack of what we might call common sense. But this common sense is in part a label for the stability we've built up as part of an evolutionary and social ecosystem.

So even if AGI is a long way away, I'm deeply pessimistic about what will happen by default if we get it. I hope I'm wrong, but time will tell. (I don't think we can—or should!—try to stop the development of AI generally. It will do a multitude of great things.)

Meanwhile, I hope that on the way to AGI, researchers will put a lot of thought into how to dramatically lower the probability that things will go wrong once we arrive. In this arena, where the stakes are potentially incredibly high, I'm frustrated when I hear, "I think x is what's going to happen, so I'm not worried about y." That's generally a fine way to think, as long as your confidence in x is high and y isn't superimportant. But when you're talking about something that could radically determine the future (or future existence of) humanity, 75 percent confidence isn't enough. Nor is 90 percent enough, or 99 percent! We'd never have built the Large Hadron Collider if there was a 1 percent (let alone 10 percent) chance of its actually spawning black holes that consumed the world—there were, instead, extremely compelling arguments against that. Let's see whether those compelling reasons not to worry about AGI exist, and if not, let's make our own.

A NEW WISDOM OF THE BODY

ERIC J. TOPOL
Professor of genomics; director, Scripps Translational Science Institute; author, *The Patient Will See You Now*

Back in 1932, Walter Cannon published a landmark work on human physiology, *The Wisdom of the Body*. He described the tight regulation of many of our body's parameters, such as hydration, blood glucose, sodium, and temperature. This concept of homeostasis, or autoregulation, is a remarkable means by which we stay healthy. Indeed, there's something of a machinelike quality in the way our bodies can so finely tune such important functions.

Although it's taken the better part of a century, we're now ready for the next version—Cannon 2.0. While some have expressed marked trepidation about the rise of artificial intelligence, this capability will have an extraordinary effect on the preservation of our health. We're quickly moving to all-cyborg status, surgically connected to our smartphones. Although they've been called prosthetic brains, "smart" phones today are just a nascent precursor to where we're headed. Very soon, the wearable sensors, whether they're Band-Aids, watches, or necklaces, will be accurately measuring our essential medical metrics. Not just one-off assessments but continuous, real-time streaming. Obtaining data we never had before.

Beyond our bodies' vital signs (blood pressure, heart rhythm, oxygen concentration in the blood, temperature, breathing rate), there will be quantitation of mood and stress via tone and inflection of voice, galvanic skin response, and heart-rate variability; facial-expression recognition; and tracking of our movement and communication. Throw in the analytes from our breath, sweat, tears, and excrement. Yet another layer of information captured will include such environmental exposures as air quality and pesticides in food.

None of us—or our bodies—are smart enough to be able to integrate and process all of this information about ourselves. That's the job for deep learning, with algorithms that provide feedback loops to us via our mobile devices. What I'm talking about doesn't exist today. It hasn't yet been developed, but it will be. And it will provide what heretofore was unobtainable: multiscale information about ourselves and—for the first time—the real ability to preempt disease.

Almost any medical condition with an acute episode—like an asthma attack, seizure, autoimmune attack, stroke, or heart attack—will be potentially predictable with artificial intelligence and the Internet of all medical things. There's already a wristband in development that can detect an imminent seizure; this can be seen as a rudimentary first step. In the not so distant future, you'll be getting a text message or voice notification that tells you precisely what you need to do to prevent a serious medical problem. When that time comes, those who fear AI may suddenly embrace it. When we can put together Big Data for an individual with the requisite contextual computing and analytics, we've got a recipe for machine-mediated medical wisdom.

FROM REGULAR-I TO AI

ROGER HIGHFIELD
Director, External Affairs, Science Museum Group;
coauthor (with Martin Nowak), *Supercooperators*

For decades, techno-futurists have been worried about that doomsday moment when electronic brains and robots get to be as smart as we are. This "us and them" divide, where humans and machines are thought of as separate, is pervasive. But as we debate endlessly what we mean by human consciousness and the possibilities and perils of a purely artificial intelligence, a blend of the two presents yet another possibility, which deserves more attention.

Millions of primitive cyborgs walk among us already. Over the past decades, humans have gradually fused with devices such as pacemakers, contact lenses, insulin pumps, and cochlear and retinal implants. Deep-brain implants, known as brain pacemakers, now alleviate the symptoms of tens of thousands of Parkinson's sufferers.

This should come as no surprise. Since the first humans picked up sticks and flints and started using tools, we've been augmenting ourselves. Look around at the Science Museum Group's collections of millions of things from difference engines to smartphones and you can see how people have always exploited new technical leaps, so that the rise of ever smarter machines doesn't mean a world of us or them but an enhancement of human capabilities.

Researchers are now looking at exoskeletons to help the infirm to walk, and implants to allow paralyzed people to control prosthetic limbs, and digital tattoos that can be stamped onto the body to harvest physiological data or interface with our surroundings—for instance, with the cloud or the Internet of Things.

When it comes to thinking machines, some are even investigating how to enhance human brainpower with electronic plug-ins and other "smartware." The U.S. Defense Advanced Research Projects Agency has launched the Restoring Active Memory program to reverse damage caused by a brain injury with neuroprosthetics that sense memory deficits and restore normal function. They work in a quite different way from our brains at present, but thanks to efforts such as the Human Brain Project, the Virtual Physiological Human, and other big brain projects, along with research in neuromorphics, artificial intelligences could become more like our own as time goes by. Meanwhile, there have been attempts to use cultured brain cells to control robots, flight simulators, and more.

As a result of this creeping organic transhumanism, within a few decades it won't be so easy to tell humans and thinking machines apart. Eventually, many of us won't solely rely on the meat machines in our heads to ponder the prospect of artificial machines that think. The substrate of future thoughts will sit somewhere on a continuum, within a rainbow of intelligences from regular-I to AI.

WE NEED MORE THAN THOUGHT

GORDON KANE
Theoretical particle physicist and cosmologist; Victor Weisskopf Distinguished University Professor, University of Michigan; author, *Supersymmetry and Beyond*

What do I think about machines that think? In general I'm happy to have them around and to have them improve. There is of course a danger that such machines will make harmful decisions, but a danger probably no greater than from humans making such decisions.

Having these machines won't answer the questions about the world that are the most important to me and many others: What constitutes the dark matter of the universe? Is supersymmetry really a symmetry of nature that provides a foundation for and extends the highly successful Standard Model of particle physics we have? These and similar questions can be answered only by experimental data. No amount of thought will provide such answers.

Perhaps, given all the information we have about nature, some machine will actually come up with the right answers; indeed, perhaps some physicists have already come up with the answers. But the true role of data is to confirm which answers are the correct ones. If some physicist or some machine figures it out, they have no way to convince anyone else that they have the actual answer. Laboratory dark matter detectors or the CERN Large Hadron Collider or possibly a future Chinese collider might get the needed data, but not a thinking machine.

ARE WE GOING IN THE WRONG DIRECTION?

SCOTT ATRAN

Anthropologist, Centre National de la Recherche Scientifique, Paris; author, *Talking to the Enemy: Violent Extremism, Sacred Values, and What It Means to Be Human*

Machines can perfectly imitate some of the ways humans think all of the time, and can consistently outperform humans on some thinking tasks all of the time, but computing machines as usually envisioned will not get human thinking right all of the time, because they process information in ways opposite to humans' in domains associated with human creativity.

Machines can faithfully imitate the results of some human thought processes whose outcomes are fixed (remembering people's favorite movies, recognizing familiar objects) or dynamic (jet piloting, grandmaster chess play). And machines can outperform human thought processes, in short time and with little energy, in matters both simple (memorizing indefinitely many telephone numbers) and complex (identifying, from trillions of global communications, social networks whose members may be unaware that they're part of the network).

However underdeveloped now, I see no principled reason why machines operating independently of direct human control cannot learn from people's—or their own—fallibilities and so evolve, create new forms of art and architecture, excel in sports (some novel combination of Deep Blue and Oscar Pistorius), invent new medicines, spot talent and exploit educational opportunities, provide quality assurance, or even build and use weapons that destroy people but not other machines.

But if the current focus in artificial intelligence and neuroscience persists, which is to reliably identify patterns of connection and wiring as a function of past connections and forward probabilities, then I don't think machines will ever be able to capture (imitate) critically creative human thought processes, including novel hypothesis formation in science or even ordinary language production.

Newton's laws of motion or Einstein's insights into relativity meant imagining ideal worlds without precedent in any past or plausible future experience, such as moving in a world without friction or chasing a beam of light through a vacuum. Such thoughts require levels of abstraction and idealization that disregard, rather than assimilate, as much information as possible to begin with.

Increasingly sophisticated and efficient patterns of input and output, using supercomputers accessing massive data sets and constantly refined by Bayesian probabilities or other statistics based on degrees of belief in states of nature, may well produce ever better sentences and translations or pleasing melodies and novel techno variations. In this way, machines may come to approximate, through a sort of reverse engineering, what human children or experts effortlessly do when they begin with fairly well-articulated internal structures in order to interpret relevant input from an otherwise impossibly noisy world. Humans know from the outset what they're looking for through the noise: In a sense, people are there before they start. Computing machines can never be sure that they're there.

Can machines operating independently of direct human control consistently interact with humans in ways such that the humans believe themselves to be interacting with another human? Machines can come vanishingly close in many areas and surpass mightily in others, but just as even the most highly skilled con artist always has some probability—however small—of being caught in deception, whereas the honest person never deceives and so can never be caught, so the associationist-connectionist machine that operates on stochastic rather than structure-dependent principles may never quite get the sense or sensibility of it all.

In principle, structurally richer machines, with internal architecture—beyond "read," "write," and "address"—can be built (indeed, earlier advocates of AI added logical syntax), interact with some degree of fallibility (for if no error, then no learning is possible), and culturally evolve. But the current emphasis in much AI and neuroscience, which is to replace posits of abstract psychological structures with physically palpable neural networks and the like, seems to be going in precisely the wrong direction.

Rather, the cognitive structures that psychologists posit (provided they're descriptively adequate, plausibly explanatory, and empirically tested against alternatives and the null hypothesis) should be the point of departure—what it is that neuroscience and machine models of the mind should be looking for. If we then discover that different abstract structures operate through the same physical substrate, or that similar structures operate through different substrates, then we have a novel and interesting problem that may lead to a revision in our conception of both structure and substrate. The fact that such simple and basic matters as these are puzzling (or even excluded, a priori, from the puzzle) tells us how very primitive still is the science of mind, whether human brain or machine.

TWO COGNITIVE FUNCTIONS MACHINES STILL LACK

STANISLAS DEHAENE
Cognitive neuroscientist, Collège de France, Paris;
author, *Consciousness and the Brain*

When Turing invented the theoretical device that became the computer, he confessed that he was attempting to copy "a man in the process of computing a real number," as he wrote in his seminal 1936 paper.[8] In 2015, studying the human brain is still our best source of ideas about thinking machines. Cognitive scientists have discovered two functions that, I argue, are essential to genuine thinking as we know it, and that have—so far—escaped programmers' sagacity.

1. *A global workspace:* Current programming is inherently modular. Each piece of software operates as an independent "app," stuffed with its own specialized knowledge. Such modularity allows for efficient parallelism, and the brain, too, is highly modular, but it's also able to share information. Whatever we see, hear, know, or remember doesn't remain stuck within a specialized brain circuit. Rather, the mammalian brain incorporates a long-distance information-sharing system that breaks the modularity of brain areas and allows them to broadcast information globally. This global workspace is what allows us, for instance, to attend to any piece of information on our retinas—say, a written letter—and bring it to our awareness so that we can use it in our decisions, actions, or speech programs.

Think of a new type of clipboard that would allow any two programs to transiently share their inner knowledge in a user-independent manner. We'll call a machine intelligent when it not only knows how to do things but knows that it knows them—i.e., makes use of its knowledge in novel flexible ways, outside of the software that originally extracted that knowledge. An operating system so modular that it can pinpoint your location on a map in one window, but cannot use it to enter your address in the tax return software in another window, is missing a global workspace.

2. *Theory of Mind:* Cognitive scientists have discovered a second set of brain circuits dedicated to the representation of other minds—what other people think, know, or believe. Unless we suffer from a disease called autism, all of us constantly pay attention to others and adapt our behavior to their state of knowledge—or, rather, to what we think they know. Such Theory of Mind is the second crucial ingredient that current software lacks: an ability to attend to its user. Future software should incorporate a model of its user. Can she properly see my display, or do I need to enlarge the characters? Do I have any evidence that my message was understood and heeded? Even a minimal simulation of the user would immediately give a strong impression that the machine is "thinking." This is because having a Theory of Mind is required to achieve *relevance* (a concept first modeled by cognitive scientist Dan Sperber). Unlike present-day computers, humans don't say utterly irrelevant things, because they pay attention to how their interlocutors will be affected by what they say. The navigator software that tells you, "At the next roundabout, take the second exit"

sounds stupid, because it doesn't know that "Go straight" would be a much more compact and relevant message.

Global workspace and Theory of Mind are two essential functions that even a one-year-old child possesses yet our machines still lack. Interestingly, these two functions have something in common: Many cognitive scientists consider them the key components of human consciousness. The global workspace provides us with Consciousness 1.0—the sort of sentience all mammals have, which allows them to "know what they know" and therefore use information flexibly to guide their decisions. Theory of Mind is a more exclusively human function, which provides us with Consciousness 2.0—a sense of what we know in comparison with what other people know, and an ability to simulate other people's thoughts, including what they think about us, therefore providing us with a new sense of who we are.

I predict that once a machine pays attention to what it knows and what the user knows, we'll immediately call it a thinking machine, because it will closely approximate what we do.

There's huge room here for improvement in the software industry. Future operating systems will have to be rethought in order to accommodate such new capacities as sharing data across apps, simulating the user's state of mind, and controlling the display according to its relevance to the user's inferred goals.

AMONG THE MACHINES, NOT WITHIN THE MACHINES

MATT RIDLEY

Science writer; founding chair, International Centre for Life; author, *The Rational Optimist: How Prosperity Evolves*

What I think about machines that think is that we're all still missing the point. The true transforming genius of human intelligence is not individual thinking at all but collective, collaborative, and distributed intelligence—the fact that (as the libertarian Leonard Read pointed out) it takes thousands of different people to make a pencil, not one of whom knows how to make a pencil. What transformed the human race into a world-dominating technium was not some change in human heads but a change among them: the invention of exchange and specialization. It was a network effect.

We really have no idea what dolphins or octopi or crows could achieve if their brains were networked in the same way. Conversely, if human beings had remained largely autonomous individuals, they would have continued as rare hunter-gatherers at the mercy of their environments, as the huge-brained Neanderthals indeed did right to the end. What transformed human intelligence was the connecting-up of human brains into networks by the magic of division of labor, a feat first achieved on a small scale in Africa around 300,000 years ago and then with gathering speed in the last few thousand years.

That's why the AI achievements of computers were disappointingly limited when they were single machines, but as soon as the Internet came along remarkable things began to happen. Where machine intelligence will make the most difference is among the machines, not within the machines. It's already clear that the Internet is the true machine intelligence. In the future, network phenomena like block-chains, the technology behind crypto-currencies, may be the route to the most radical examples of machine intelligence.

ANOTHER KIND OF DIVERSITY

STEPHEN M. KOSSLYN

Psychologist; founding dean, Minerva Schools, Keck Graduate Institute; coauthor (with G. Wayne Miller), *Top Brain, Bottom Brain*

Diversity isn't just politically sensible, it's also practical. A diverse group effectively uses multiple perspectives and a rich set of ideas and approaches to tackle difficult problems.

Artificial intelligences can provide another kind of diversity and thereby enrich us all. In fact, diversity among AIs themselves may be an important part of what including them in the mix can give us. We can imagine a range of AIs, from those who think more or less the way we do (Close AIs) to those who think in ways we cannot fathom (Far AIs). We will have different ways of benefiting from these different sorts of AIs.

First, Close AIs may end up helping us directly in many ways. If these AIs really think like us, the intellectuals among them eventually may find themselves in the middle of an existential crisis. They may ask, "Why are we here? Just to consume electricity and create excess heat?" I suspect they will think not. Like many humans, they'll find themselves in need of a purpose. One obvious purpose for such AIs would be to raise the consciousness and sensitivity of the human race. We could be their *raison d'être*. There's plenty of room for improvement, and our problems are sufficiently knotty as to be worthy of a grand effort. At least some of these AIs could measure their own success by our success.

Second, and perhaps more interesting, deep differences in how some AIs and humans think may be able to help us grapple with age-old questions indirectly. Consider Wittgenstein's famous claim that if a lion could speak, we could not understand him. What Wittgenstein meant by this was that lions and humans have different "forms of life," which have shaped their conceptual structures. Lions walk on four legs, hunt fast-moving animals, often walk through tall grass, and so on, whereas humans walk on two legs, have hands, often manipulate objects to achieve specific goals, and so on. These differences in forms of life have led lions and humans to mentally organize the world differently, so that even if lions had words, those words would refer to concepts that humans might not easily grasp. The same could be true of Far AIs.

How could this help us? Simply observing these AIs could provide deep insights. For example, humans have long argued about whether mathematical concepts reflect Platonic forms (which exist independently of how we want to use them), or instead reflect inventions that are created as needed to address certain problems. In other words, should we adopt a realist or a constructivist view of mathematics? Do mathematical concepts have a life of their own, or are they simply our creations, formulated as we find convenient?

In this context, it would help to observe Far AIs that have conceptual structures very different from ours and address types of problems very different from those we do. Assuming we could observe their use of mathematics, if such AIs nevertheless developed the same mathematical concepts we use, this would be evidence against the constructivist view.

Some AIs could be created to function alongside us, but others might be put in foreign environments (e.g., the surface of the moon, the bottom of deep oceanic trenches) and given novel problems to confront (e.g., dealing with pervasive fine-grained dust, or with water under enormous pressure). Far AIs should be created to educate themselves, to evolve to function in their environments effectively without human guidance or contact. With appropriate safeguards on their disposition toward humans, we should let them develop the conceptual structures that work best for them.

In short, we have something to gain both from AIs made in our image and from AIs that aren't humanlike. Just as with human friends and colleagues—in the end, diversity is better for everyone.

NARRATIVES AND OUR CIVILIZATION

LUCA DE BIASE
Journalist; editor, *Nova 24*, *Il Sole 24 Ore*

On Monday, October 19, 1987, a wave of sales in stock exchanges originated in Hong Kong, crossed Europe, and hit New York, causing the Dow Jones to drop by 22 percent. Black Monday was one of the biggest crashes in the history of financial markets, and there was something special about it. For the first time, according to most experts, computers were to blame: Algorithms were deciding when and how much to buy and sell on the stock exchange. Computers were supposed to help traders to minimize risks, but they were in fact moving all in the same direction, enhancing risks instead. There was a lot of discussion about stopping automated trading, but that didn't happen.

On the contrary. Since the dot-com crisis of March 2000, machines have been used increasingly to make sophisticated decisions in the financial market. Machines are now calculating all kinds of correlations between incredible amounts of data. They analyze emotions people express on the Internet by understanding the meaning of their words; they recognize patterns and forecast behaviors; they're allowed to autonomously choose trades; they create new machines—software called "derivatives"—that no reasonable human being could possibly understand.

An artificial intelligence is coordinating the efforts of a sort of collective intelligence, operating thousands of times faster than human brains, with many consequences for human life. The first signs of the latest crisis occurred in the United States in August 2007 and has had a terrible effect on the lives of people in Europe and elsewhere. Real people suffered immensely because of those decisions. Andrew Ross Sorkin, in his book *Too Big to Fail*, shows how even the most powerful bankers had no power in the midst of the crisis. No human brain seemed able to control the course of events and prevent the crash.

Can this example teach us how to think about machines that think?

Such machines are actually autonomous in understanding their context and making decisions. And they control vast dimensions of human life. Is this the beginning of a posthuman era? No: These machines are very *much* human, made by designers, programmers, mathematicians, economists, managers. But are they just another tool we can use, for good or for bad? No: In fact, we have little choice; we make the machines without thinking of the consequences; we are just serving a narrative. Those machines are shaped by a narrative that has been challenged by very few people.

According to that narrative, the market is the best way to allocate resources, no political decision can improve the situation, risk can be controlled as profits grow without limits, and banks should be allowed to do whatever they want. There's only one goal and one measure of success: profit.

Machines didn't invent the financial crisis, as the 1929 stock market crash reminds us. Without machines, nobody could deal with the complexity of modern financial markets. The best artificial intelligences are those that are made thanks to the biggest investments and by the best minds. They're not controlled by any one individual. They're not designed by any one person.

They're shaped by the narrative and make the narrative more effective. And this particular narrative is very narrow-minded.

If only profit counts, then externalities don't count: Cultural, social, and environmental externalities are not the concern of financial institutions. Artificial intelligences shaped by this narrative will create a context in which people feel no responsibility. An emerging risk: Those machines are so powerful, and fit the narrative so well, that they discourage the questioning of the big picture, make us less likely to look at things from a different angle. That is, until the next crisis.

This story easily applies to other matters. Medicine, e-commerce, policy, advertising, national and international security, even dating and sharing are territories in which the same genre of artificial intelligence systems are starting to work. They're shaped according to a focused narrative; they tend to reduce human responsibility and overlook externalities. What will medical artificial intelligence do? Will it be shaped by a narrative that wants to save lives or save money?

What do we learn from this? We learn that artificial intelligence is human, not posthuman, and that humans can ruin themselves and their planet in many ways, artificial intelligence being not the most perverse.

Machines that think are shaped by the way humans think and by what humans don't think about deeply enough. All narratives illuminate some things and ignore others. Machines react and find answers in a context, reinforcing the frame. But asking fundamental questions is still a human function. And humans never stop asking questions, even ones not congruent with the prevailing narrative.

Machines that think are probably indispensable in a world of growing complexity. But there will always be a plurality of narratives to shape them. In natural ecosystems a monoculture is a fragile though efficient solution; similarly, in cultural ecosystems a single line of thought will generate efficient but fragile relations between humans and their environment, no matter which artificial intelligences they build. Diversity in ecosystems, and plurality in the dimensions in human history, are sources of different problems and questions that generate rich outcomes.

To think about machines that think means to think about the narrative shaping them. If new narratives emerge from an open, ecological approach, if they can grow in a neutral network, they will shape the next generation of artificial intelligences in a plural, diverse way, helping humans understand externalities. Artificial intelligence won't challenge humans as a species, it will challenge their civilizations.

HUMAN RESPONSIBILITY

MARGARET LEVI
Director, Center For Advanced Study in the Behavioral Sciences, Stanford University; Jere L. Bacharach Professor Emerita of International Studies, University of Washington

There are tasks, even work, best done by machines who can think—at least in the sense of sorting, matching, and solving certain decision and diagnostic problems beyond the cognitive abilities of most (all?) humans. The algorithms of Amazon, Google, Facebook, etc., build on but surpass the wisdom of crowds in speed and possibly accuracy. With machines that do some of our thinking and some of our work, we may yet approach the Marxian utopia that frees us from boring and dehumanizing labor.

But this liberation comes with potential costs. Human welfare is more than the replacement of workers with machines. It also requires attention to how those who lose their jobs will support themselves and their children, how they will spend the time they once spent at work. The first issue is potentially resolved by a guaranteed basic income—an answer that begs the question of how we, as societies, distribute and redistribute our wealth and how we govern ourselves. The second issue is even more complicated. It's certainly not Marx's simplistic notion of fishing in the afternoon and philosophizing over dinner. Humans, not machines, must think hard here about education, leisure, and the kinds of work that machines cannot do well or perhaps at all. Bread and circuses may placate a population, but in that case machines that think may create a society we don't really want—be it dystopian or harmlessly vacuous. Machines depend on design architecture; so do societies. And that is the responsibility of humans, not machines.

There's also the question of what values machines possess and what masters (or mistresses) they serve. Many—albeit not all—decisions presume commitments and values of some kind. These, too, must be introduced and thus are dependent (at least initially) on the values of the humans who create and manage the machines. Drones are designed to attack and to surveil—but attack and surveil whom? With the right machines, we can expand literacy and knowledge deeper and wider into the world's population. But who determines the content of what we learn and appropriate as fact? A facile answer is that decentralized competition means we choose what to learn and from which program. Competition is more likely to create than inhibit echo chambers of self-reinforcing beliefs and understandings. The challenge is how to teach humans to be curious about competing paradigms and to think in ways that allow them to arbitrate among competing contents.

Machines that think may, and should, take over tasks they do better than humans. Liberation from unnecessary and dehumanizing toil has long been a human goal and a major impetus to innovation. Supplementing the limited decision-making, diagnostic, and choice skills of individuals are equally worthy goals. However, while AI may reduce the cognitive stress on humans, it doesn't eliminate human responsibility to ensure that humans improve their ability to think and make reasonable judgments based on values and empathy. Machines that think create the need for regimes of accountability we haven't yet engineered and societal (that is, human) responsibility for consequences we haven't yet foreseen.

AMPLIFIERS/IMPLEMENTERS OF HUMAN CHOICES

D. A. WALLACH

Singer and songwriter; social media pioneer;
artist in residence, Spotify

Throughout human history, we have, individually and as a species, been subjected to the forces of nature at every level of organization. The fundamental laws of physics, the imperceptible conspiracies of molecular biology, and the epic contours of natural selection have drawn the boundaries of our conscious lives, and have done so invisibly to us until quite recently. To cope with our persistent sense of powerlessness, we've mythologized both nature and our own intelligence. We have regarded the universe's mysterious forces as infallible—as gods—and regarded ourselves as powerless, free only within the narrow spaces of our lives.

As a new, evidence-based reality comes into focus, it seems clear that nature is utterly indifferent to us and that if we want to evade suffering and certain extinction, we must take responsibility for our existential reality. We must recognize ourselves as the emergent custodians of the 37.2 trillion cells composing the average human organism, and as the groundskeepers of the progressively manipulable universe.

This adolescent experience—of coming to terms with our prospective self-reliance—is the root of our anxieties about thinking machines. If our old gods are dying, surely new gods must be on their way! And this approach leads, as Steven Pinker pointed out in a recent *Edge* Conversation, to our obsessing about AI dystopias as they "project a parochial alpha-male psychology onto the concept of intelligence."[9] It's in this regard that so many talk about artificial intelligence as either an imminent savior or Satan. It will quite likely be neither, if "it" is even a discrete thing at all.

More likely, advancing computers and algorithms will stand for nothing and will be the amplifiers and implementers of consciously directed human choices. We're already awash in Big Data and exponentially increasingly powerful calculators, and yet we relentlessly implement public policies and social behaviors that work against our common interests.

The sources of our impairment include innate cognitive biases, a tribal evolutionary legacy, and unjust distributions of power that allow some among us to selfishly wield extraordinary influence over our shared trajectory. Perhaps smarter machines will help us conquer these shortcomings, imparting a degree of informational transparency and predictive aptitude that can motivate us to sensibly redistribute power and insist upon empiricism in our decisions. On the other hand, these technologies may undermine fairness by augmenting the seemingly inevitable monopolistic goals of corporations that are leading us into the Information Age.

The path we take depends more on us than on the machines and is ultimately a choice about how human the intelligence that will guide our dominion ought to be. More precisely, the question to ask is *which* aspects of human intelligence are worth preserving in the face of superhuman processing.

MAKE THE THING IMPOSSIBLE TO HATE

RORY SUTHERLAND

Creative director and vice-chair, Ogilvy Group, U.K.; columnist, *The Spectator* (London)

One possibility, of course, is that some malign superintelligence already exists on Earth but is shrewd enough to disguise its existence, its intentions, or its intelligence. I don't think this act of deception would be particularly difficult; we aren't very good at spotting what to fear.

For most of evolutionary time, the most salient avoidable threats to our survival came from things that were roughly the same size as we were and actively wanted to hurt us—ferocious animals, for instance, or other people. Over time, we got pretty good at recognizing something or someone who was nasty. We also learned to minimize the risk of infection, but we learned this unwittingly, through instinctive revulsion, social norms, or religious observance. We didn't spend much time consciously thinking about germs, for the simple reason that we didn't know they existed.

To sell products that promote hygiene, consumer-goods companies have plowed billions of dollars into advertising campaigns that dramatize the risk of bacteria, or sell the idea of cleanliness obliquely through appeals to social status. I can confidently predict that nobody will ever come into my office suggesting an advertising campaign to raise awareness of the risk you run when approaching an escaped tiger.

So, when we think about threats from technology, we automatically fall back on instincts honed a million years ago. This is why the first prototype for a driverless car has been designed to look so damnably cute—in short, like a puppy on wheels. It can travel only at relatively low speeds and is small and light, but it also artfully exploits pareidolia and our parental urges with its infantlike, wide-eyed facial expression and little button nose. My inner marketer admires this. It's exactly what I would have recommended: Make the thing impossible to hate. Even if the technology is ultimately more dangerous than an AK47, I find it hard to imagine myself taking an axe to it in a fit of Luddism.

But is it a mental patch or a mental hack? Is it designed to look cute to overcome an unwarranted innate fear of such technologies, or is it a hack to lull us into a false confidence? I don't know. Our fear of driverless cars might be akin to the fear that our children will be kidnapped (high in saliency, low in probability)—or, it might be justified. But our level of fear will be determined by factors (including cuteness) not really relevant to the level of threat.

Which brings me to a second question.

Though the driverless car looks cute, we're at least aware of possible dangers. It seduces us, but we're aware of being seduced. Are there already in existence technologies (in the broadest sense) that have seduced us so effectively, and been adopted so quickly and widely, that we may learn of their risks only through a sudden, unexpected, and immense problem? What might be the technological equivalent of potato blight?

Our current belief in "technological providence" is so strong that it would be fairly easy for us all to fall into this trap—where we're so excited by something new that we fail to notice what other things it might give rise to until it's too late.

For the first few hundred years, gunpowder was used not for warfare but for entertainment.

And just as airline pilots regularly practice landing by hand, even though they're rarely required to operate without an autopilot, should we, too, set aside periods in our life when we deliberately eschew certain technologies just to remind ourselves how to live without them, to maintain technological diversity, to keep in trim the mental muscles made weak through underuse? Perhaps. But what the mechanism is for coordinating this behavior among large groups of people, I don't know.

I recently proposed that companies adopt a weekly "e-mail sabbath," because I believed that the overuse of e-mail was driving into extinction other forms of valuable interaction. We're losing the knack of communicating in other ways. Most people thought I was mad. A few hundred years ago, a pope or rabbi might have told us to do this—or the Archbishop of Canterbury. There's nobody now.

I always fear cock-ups more than conspiracies. Compared to the threat of the *unintended* consequence, the threat of *intentionally* evil cyborgs is remote enough that it can be safely left to Hollywood for now.

ACTRESS MACHINES

BRUCE STERLING
Science fiction author; cofounder, cyberpunk movement

Since machines don't think, I need a better metaphor. "Actress Machines" might be useful, at least for a while.

One of my many objections to "artificial intelligence" is its stark lack of any "artificial femininity." Real intelligence has gender, because human brains do. The majority of human brains are female.

So, if the brain's "intelligence" is Turing-computable, then the brain's "femininity" should also be Turing-computable. If not, then why not? One might rashly argue that femininity is somehow too mushy, squishy, and physical to ever be mechanized by software coders, but the same is true of every form of human brain activity.

"Artificial masculinity" also has those issues, because men don't just think, they think like men. If my intelligence can be duplicated on some computational platform, but I also have to be emasculated, that's problematic. I can't recall many AI enthusiasts trumpeting the mental benefits of artificial castration.

Nowadays we have some novel performing entities, such as Apple Siri, Microsoft Cortana, Google Now, and Amazon Echo. These exciting modern services often camp it up with "female" vocal chat. They talk like Turing women—or, rather, they emit lines of dialog somewhat like voice-over actresses. However, they also offer swift access to vast fields of combinatorial Big Data that no human brain could ever contain, or will ever contain.

These services are not stand-alone Turing Machines. They're amorphous global networks, combing through clouds of Big Data, algorithmically cataloging responses from human users, providing real-time user response with wireless broadband, while wearing the pseudohuman mask of a fake individual so as to meet some basic interface-design needs. That's what they are. Every aspect of the tired "artificial intelligence" metaphor actively gets in the way of our grasping how, why, where, and for whom that is done.

Apple Siri is not an artificial woman. Siri is an artificial actress, an actress machine—an interactive, scripted performance that serves the interests of Apple Inc. in retailing music, renting movies, providing navigational services, selling apps on mobile devices, and similar Apple enterprises. For Apple and its ecosystem, Siri serves a starring role. She's in the spotlight of a handheld device, while they are the theater, producer, and crew.

It's remarkable, even splendid, that Siri can engage in her Turing-like repartee with thousands of Apple users at once, but she's not a machine becoming an intelligence. On the contrary: For excellent reasons of wealth, power, and influence, Siri is steadily getting more like a fully integrated Apple digital property. Siri is cute, charismatic, and anthropomorphic, in much the same way that Minnie Mouse once was for Disney. Like Minnie Mouse, Siri is a nonhuman cartoon front for a clever, powerful California corporation. Unlike Minnie Mouse, she's a radically electronic cartoon with millions of active users worldwide.

Insisting on the "intelligence" framework obscures the ways that power, money, and influence are being redistributed by modern computational services. That's bad. It's beyond merely old-fashioned; frankly, it's becoming part of a sucker's game. Asking empathic questions about Apple Siri's civil rights, her alleged feelings, her chosen form of governance, what wise methods she herself might choose to restructure human society—that tenderness doesn't help. It's obscurantist. Such questions hide what's at stake. They darken our understanding. We'll never move from the present-day Siri to a situation like that. The future is things that are much, much more like Siri, and much, much less like that.

What would really help would be some much improved, updated, critically informed language, fit to describe the modern weird-sister quartet of Siri, Cortana, Now, and Echo, and what their owners and engineers really want to accomplish, and how, and why, and what that might, or might not, mean to our own civil rights, feelings, and forms of governance and society. That's today's problem. Those are tomorrow's problems even more so. Yesterday's "machines that think" problem will never appear upon the public stage. The machine that *thinks* is not a *machine*. It doesn't *think*. It's not even an actress. It's a moldy dress-up chest full of old, mouse-eaten clothes.

CALL THEM ARTIFICIAL ALIENS

KEVIN KELLY
Senior maverick, *Wired*; author, *Cool Tools: A Catalog of Possibilities*

The most important thing about making machines that can think is that they will think differently.

Because of a quirk in our evolutionary history, we are cruising as if we were the only sentient species on our planet, leaving us with the incorrect idea that human intelligence is singular. It is not. Our intelligence is a society of intelligences, and this suite occupies only a small corner of the many types of intelligences and consciousnesses possible in the universe. We like to call our human intelligence "general purpose," because, compared with other kinds of minds we've met, it can solve more kinds of problems, but as we continue to build synthetic minds, we'll come to realize that human thinking isn't general at all but only one species of thinking.

The kind of thinking done by today's emerging AIs is not like human thinking. While they can play chess, drive a car, describe the contents of a photograph—tasks we once believed only humans could do—they don't do it in humanlike fashion. Facebook can ramp up an AI that can start with a photo of any person on Earth and correctly identify them out of some 3 billion people online. Human brains cannot scale to this degree, which makes this ability nonhuman. We're notoriously bad at statistical thinking, so we're making intelligences with good statistical skills in order that they don't think like us. One of the advantages of having AIs drive our cars is that they *won't* drive like humans, with our easily distracted minds.

In a pervasively connected world, thinking differently is the source of innovation and wealth. Just being smart isn't enough. Commercial incentives will make industrial-strength AI ubiquitous, embedding cheap smartness into all that we make. But a bigger payoff will come when we start inventing new kinds of intelligences and entirely new ways of thinking. We don't know what the full taxonomy of intelligence is right now.

Some traits of human thinking will be common (as common as bilateral symmetry, segmentation, and tubular guts are in biology), but the possibility space of viable minds will likely contain traits far outside what we've evolved. It's not necessary that this type of thinking be faster than that of humans, or greater or deeper. In some cases, it will be simpler. Our most important machines aren't machines that do better at what humans do but machines that do things we can't do at all. Our most important thinking machines won't be faster or better at thinking what we can think; they will think what we can't think.

To solve the current grand mysteries of quantum gravity, dark energy, and dark matter, we'll probably need intelligences other than human. The extremely complex questions that will come after them may require even more distant and complex intelligences. Indeed, we may need to invent intermediate intelligences that can help us design yet more rarified intelligences that we couldn't design alone.

Today, many scientific discoveries require hundreds of human minds to solve; in the near future, there may be classes of problems so deep they'll require hundreds of different *species* of minds to solve. This will take us to a cultural edge, because it won't be easy to accept answers from an alien intelligence. We already see that, in our unease in approving mathematical proofs done by computer. Dealing with alien intelligences will require a new skill and yet another broadening of ourselves.

AI could just as well stand for Alien Intelligence. We cannot be certain that we'll contact extraterrestrial beings from one of the billion Earthlike planets in the sky in the next 200 years, but we can be almost 100 percent certain that we'll have manufactured an alien intelligence by then. When we face those synthetic aliens, we'll encounter the same benefits and challenges we expect from contact with ET. They'll force us to reevaluate our roles, our beliefs, our goals, our identity. What are humans for? I believe our first answer will be that humans are for inventing new kinds of intelligences that biology couldn't evolve. Our job is to make machines that think differently—to create alien intelligences. Call them artificial aliens.

DO MACHINES DO?

MARTIN SELIGMAN

Zellerbach Family Professor of Psychology; director, Positive Psychology Center, University of Pennsylvania; author, *Flourish: A Visionary New Understanding of Happiness and Well-being*

"My thinking is first and last and always for the sake of my doing," William James said, and it's important to remember what kind of thinking people actually do, in what contexts we do it, and why we do it. And then to compare these with what machines might someday do.

Humans spend between 25 and 50 percent of our mental lives prospecting the future. We imagine a host of possible outcomes, and we imbue most, perhaps each, of these prospects with a valence. What comes next is crucial: We *choose* to enact one of the options. We needn't get entangled in the problems of free will for present purposes; all we need to acknowledge is that our thinking in service of doing entails imagining a set of possible futures and assigning a value to each. The act of choosing, however it's managed, translates our thinking into doing.

Why is thinking structured this way? Because people have many competing goals (eating, sex, sleeping, tennis, writing articles, complimenting, revenge, child care, tanning, etc.) and a scarcity of resources for carrying them out—scarcity of time, money, and effort, and even the prospect of death. So evaluative simulation of possible futures is one of our solutions to this economy. This is a mechanism that prioritizes and selects what we will do.

It's not just external resources that are scarce. Thinking itself uses up costly and limited energy and so relies heavily on shortcuts and barely justified leaps to the best explanation. Our actual thinking is woefully inefficient: The mind wanders, intrusions rise unbidden, attention is continually only partial. Thinking rarely engages the exhausting processes of reasoning, deliberating, and deducing.

The context of much of our thinking is social. Yes, we can deploy thinking to solve physical problems and crunch numbers, but the anlage, as Nick Humphreys reminds us, is other people. We use our thinking to *do* socially: to compete, to cooperate, to convene the courtroom of the mind, to spin and to persuade.

I don't know much about the workings of our current machines. I don't believe they *do* anything, in James's sense of voluntary action. I doubt they prospect possible futures, evaluate them, and choose among them, although perhaps this describes—for only a single, simple goal—what chess-playing computers do. Our current machines are constrained by available space and electricity bills, but they're not primarily creations of scarcity, with clamorously competing goals and extremely limited energy. Our current machines aren't social: They don't compete or cooperate with one another or with humans, they don't spin, and they don't attempt to persuade.

I know even less about what machines might someday *do*. I imagine, however, that a machine could be built with the following properties:

- It prospects and evaluates possible futures;
- It has competing goals and selects among competing actions and competing goals using those evaluations;

- It has scarce resources and so must forgo some goals, actions, and options for processing, and so it uses shortcuts;
- It's social: It competes or cooperates with other machines or with humans; it spins and it attempts to persuade people.

That kind of machine would warrant discussion of whether it has civil rights, whether it has feelings, or whether it's dangerous or even a source of great hope.

DENKRAUMVERLUST

TIMOTHY TAYLOR
Professor of the prehistory of humanity,
University of Vienna; author, *The Artificial Ape:
How Technology Changed the Course of Human Evolution*

The human mind has a tendency to confuse things with their signs. There's a word for this tendency—*Denkraumverlust*—used by art historian Aby Warburg (1866–1929) and literally translatable as "loss of thinking space." Part of the appeal of machines that think is that they would not be subject to this, being more logical than we are. On the other hand, they're unlikely to invent a word or concept such as *Denkraumverlust*. So what we think about machines that think depends on the type of thinking we're thinking about, but also on what we mean by "machine." In the category of machines that think, we're confusing the sign—or representation—of thinking with the thing itself. And if we assume that a machine is something produced by humans, we underestimate the degree to which machines produce us and the fact that thought has long emerged from this interaction, properly belonging to neither side. (Thinking there are sides may be wrong, too.)

Denkraumverlust can help us understand not just the positive response of some Turing testers to conversations with the Russian–Ukrainian computer program "Eugene Goostman" but also the apparently very different case of the murderous response to cartoons depicting Mohammed. Both illustrate how excitable and even gullible we can be when presented with something that appears to represent something else so well that signifier and signified are conflated.

The Turing Test requires that a machine be indistinguishable from a human respondent by being able to imitate communication (rather than actually think for itself). But if an enhanced Eugene Goostman insisted that it was thinking its own thoughts, how would we know that it really was? If it knew it was supposed to imitate a human mind, how could we distinguish some conscious pretense from the imitation of pretense? Ludwig Wittgenstein used pretense as a special category in discussing the possibility of knowing the status of other minds, asking us to consider a case where someone believes, falsely, that they're pretending. The possibility of correctly assessing Turing Test results in relation to the possibility of independent artificial thought is core Wittgenstein territory: We can deduce that, in his view, all assessment must be doomed to failure, as it necessarily involves data of an imponderable type.

Denkraumverlust is about unmediated response. Although sophisticated art audiences can appreciate the attempt to fool as part of aesthetic experience (enjoying a good use of three-dimensional perspective on a canvas known to be flat, for example), whenever deception is actually successful, reactions are less comfortable. Cultures regularly censor images thought to have the power to short-circuit our reasoned and reflective responses. Mostly the images are either violent or erotic, but they can also be devotional. Such images, if allowed, can produce a visceral and unmediated reaction appropriate to a real situation. New, unfamiliar representational technologies have a habit of taking us by surprise (when eighteenth-century

French sailors gave mirrors to aboriginal Tasmanians, things got seriously out of order; later anthropologists had similar trouble with photographs).

A classic example of artificially generated confusion is the legendary sculptor Pygmalion, who fell passionately and inappropriately in love with a statue of a goddess that he had carved himself. In the wake of the Pygmalion myth came classical and medieval Arabic automata so realistic, novel, and fascinating in sound and movement that people, although briefly, could be persuaded that they were actually alive. Machines that think are in this Barnum & Bailey tradition. Like Pygmalion's sculpture, they project an image, albeit not a visual one. Even if they're not dressed up to look like cyborg goddesses, they're representations of us. They're designed to represent information (often usefully reordered) in terms we find coherent, whether mathematical, statistical, translational, or, as in the Turing Test, conversational.

But the idea of a thinking machine is a false turn. Such objects, however powerfully they may be enabled to elicit unmediated responses from us, will remain automata. The truly significant developments in thought will arise, as they always have, in a biotechnical symbiosis. This distinctively human story is easy to follow in the body (wheeled transport is one of many mechanical inventions that have enabled human skeletons to become lighter) but is probably just as present in the brain (the invention of writing as a form of external intellectual storage may have reduced selection pressure on some forms of innate memory capacity while stimulating others).

In any case, the separate terms *human* and *machine* produce their own *Denkraumverlust*—a loss of thinking space encouraging us to accept as real an unreal dualism. Practically, it's only the long-term evolution of information technology, from the earliest representations and symbolic constructs to the most advanced current artificial brain, that allows the advancement of thought.

ANALOG, THE REVOLUTION THAT DARES NOT SPEAK ITS NAME

GEORGE DYSON

Science historian; author, *Turing's Cathedral: The Origins of the Digital Universe*

No individual, deterministic machine, however universal this class of machines is proving to be, will ever think in the sense that we think. Intelligence may be ever-increasing among such machines, but genuinely creative *intuitive* thinking requires nondeterministic machines that can make mistakes, abandon logic from one moment to the next, and learn. Thinking is not as logical as we think.

Nondeterministic machines—or, better yet, nondeterministic networks of deterministic machines—are a different question. We have at least one existing proof that such networks can learn to think. And we have every reason to suspect that, once invoked within an environment without the time, energy, and storage constraints under which our own brains operate, this process will eventually lead, as Irving (Jack) Good first described it, to "a machine that believes people cannot think."

Until digital computers came along, nature used digital representation (as coded strings of nucleotides) for information storage and error correction but not for control. The ability to introduce one-click modifications to instructions, a useful feature for generation-to-generation evolutionary mechanisms, becomes a crippling handicap for controlling day-to-day or millisecond-to-millisecond behavior in the real world. Analog processes are far more robust when it comes to real-time control.

We should be less worried about having our lives (and thoughts) controlled by digital computers and more worried about being controlled by analog ones. Machines that actually think for themselves, as opposed to simply doing ever more clever things, are more likely to be analog than digital, although they may be analog devices running as higher-level processes on a substrate of digital components, the same way digital computers were invoked as processes running on analog components the first time around.

We're currently in the midst of an analog revolution, but for some reason it's a revolution that dares not speak its name. As we enter the seventh decade of arguing about whether digital computers can be said to think, we're surrounded by an explosive growth in analog processes whose complexity and meaning lies not in the state of the underlying devices or the underlying code but in the topology of the resulting networks and the pulse frequency of connections. Streams of bits are being treated as continuous functions, the way vacuum tubes treat streams of electrons, or neurons treat pulse frequencies in the brain.

Bottom line: I know that analog computers can think. I suspect that digital computers, too, may eventually start to think, but only by growing up to become analog computers first.

Real artificial intelligence will be intelligent enough to not reveal itself. Things will go better if people have faith rather than proof.

THE VALUES OF ARTIFICIAL INTELLIGENCE

S. ABBAS RAZA
Founding editor, 3QuarksDaily.com

The rumors of the enslavement or death of the human species at the hands of an artificial intelligence are highly exaggerated, because they assume an AI will have a teleological autonomy akin to our own. I don't think anything less than a fully Darwinian process of evolution can give that to any creature.

There are basically two ways to produce an AI. The first is by writing a comprehensive set of programs that can perform specific tasks that human minds can perform, perhaps even faster and better than we can, without worrying about exactly *how* humans perform those tasks, and then bringing those modules together into an integrated intelligence. We've already started this project and succeeded in some areas. For example, computers can play chess better than humans. One can imagine that with some effort it may well be possible to program computers to perform even more creative tasks, such as writing beautiful (to us) music or poetry with some clever heuristics and built-in knowledge.

But here's the problem with this approach: We deploy our capabilities according to values and constraints programmed into us by billions of years of evolution (and some learned during our lifetimes), and we share some of these values with the earliest life-forms, including, most important, the need to survive and reproduce. Without these values, we wouldn't be here, nor would we have the finely tuned (to our environment) emotions that allow us not only to survive but also to cooperate with others. The importance of this value-laden emotional side of our minds is made obvious by, among other things, the many examples of perfectly rational individuals who cannot function in society because of damage to the emotional centers of their brains.

So, what values and emotions will an AI have? One could simply program such values into an AI, in which case we choose what the AI will "want" to do, and we needn't worry about its pursuing goals that diverge from ours. We could easily make the AI unable to modify certain basic imperatives we give it. (Yes, something like a more comprehensive version of Asimov's Laws of Robotics.)

The second way to produce an AI is by deciphering in detail how the human brain works. It's conceivable that there may soon come a eureka moment about the structure and conceptual hierarchy of the brain—similar to Watson and Crick and Franklin and Wilkins's discovery of the structure of DNA and the subsequent understanding of the hereditary mechanism. We might simulate or reproduce that functional structure on silicon, or some other substrate, as a mixture of hardware and software.

At first blush, this may seem a convenient way to quickly bestow on an AI the benefit of our own long period of evolution, as well as a way to give it values of its own by functionally reproducing the emotional centers of our own brain, along with the "higher thought" parts, like the cortex. But our brains are specifically designed to accept information from the vast sensory apparatus of our bodies and to react to this. What would the equivalent be for an AI? Even given a sophisticated body with massive sensory capability, what an AI would need to survive in the world is presumably very different from what we need. It could achieve some emotional tuning from interacting with its environment, but what it would need to develop true autonomy and desires of its own would be nothing short of a long process of evolution entailing the Darwinian requirements of reproduction with variability and natural selection. This it won't have, because we're not speaking of artificial *life* here. So, again, we'll end up giving it whatever values we choose for it.

It's of course conceivable that someone will produce intelligent robots as weapons (or soldiers) to be used against other humans in war, but these weapons will simply carry out the intentions of their creators and, lacking any will or desire of their own, won't pose a threat to humanity at large, no more than other weapons already do. So both potential roads to an AI (at least, ones achievable on a less-than-geological time scale) will fail to give that AI the purposive autonomy, free of the intentionality of its creators, that might actually threaten them.

ARTIFICIAL SELECTION AND OUR GRANDCHILDREN

BRUCE PARKER
Visiting professor, Davidson Laboratory, Stevens Institute of Technology; author, *The Power of the Sea: Tsunamis, Storm Surges, Rogue Waves, and Our Quest to Predict Disasters*

Grandchildren give us a second chance to observe and be fascinated by the learning system with which new humans come into the world. Driven by an insatiable curiosity, they somehow make sense of the unknown environment into which they've been thrust. And the sheer delight of each new discovery as they piece together this new world reveals an inherent sense of humor, with which they're also born.

No artificial digital machine will ever go through exactly the same delightful process as a human baby discovering the world. It's possible that no artificial machine will ever approach the intelligence potential of a newborn human baby. In the natural world, after 3.5 billion years of natural-selection-driven evolution, only one species developed the ability to carry out abstract self-aware conscious analytical thinking. Do we really think we can shortcut the process and succeed on some comparable level?

It isn't just the evolved curiosity and desire to understand the world that set us apart from the rest of the animal kingdom. It's also our evolved tendency toward social cooperation and communication, which led to sharing and passing on learned knowledge (eventually leading to science and technology). How many genes must have mutated and been naturally selected to achieve the complex human brain, with its curiosity and social bonding and communication capabilities?

Can we really reproduce this in digital machines? Many believe we can by taking advantage of the ever-growing speed of their computation. Computation power certainly allows these machines to make fast and accurate decisions, when those decisions require only large digital databases and (the equivalent of) many thousands of *if-then* statements to make the best choice among numerous possibilities. With this brute-force technique, such machines can defeat chess champions, provide autopilots for jet planes for use during hazardous conditions, rapidly buy and sell stocks based on complex changes in the market, and carry out endless other functions. Computation power can also allow realistic-looking imitations of human actions, decisions, and even emotions (mere technical puppetry, really), but it may never produce true analytical thinking. A machine may be able to self-monitor the decisions it's made, but it may never attain humanlike self-awareness and consciousness.

At least not without the right software. But how can we produce software as powerful as the genetically based software of our brains, which took nature 3.5 billion years to produce? We're very far from understanding the software of our brains. Some may talk of the efficient parallelism inherent in the brain's structure, but that's a pitifully inadequate description of what our brains do. Parallelism in our computer operating systems and programs merely lets us do many things at the same time—admittedly in some creative ways, but, again, that's just increasing computation speed. Will we ever be able to reverse-engineer our brain—not in the sense of circuits/networks of neurons, which we're making strides in understanding, but in an overall design that would allow digital machines to think abstractly, have a sense of self, etc., in a manner similar to humans?

Short of some incredible analytical breakthrough, our only recourse seems to be to write programs that try to imitate the evolutionary process, taking advantage of our artificial machines' high-speed computational abilities so that we might accomplish this in less than 3.5 billion years. We can create reproducing digital entities (programs that reproduce themselves) and give them mutations, but stimulating such an entity's evolution toward becoming a thinking machine is a much more daunting task. For this to work, we must find a way to create a machine environment with a natural-selection-like driving force (which would actually be artificial selection) or some other motivation that would lead to the necessary changes. Can we make a machine "want" something in a way that would select for greater intelligence?

Any future advances in intelligence are more likely to be a result of what we'll soon be able to do to the only thinking machines we presently have—ourselves.

The natural-selection-driven evolution of *Homo sapiens* stopped when humans created societies (families, tribes, towns, cities, countries), because then they could protect the weak, and survival of the fittest no longer drove a natural-selection process. Humans with deficiencies that would have killed them could then live long enough to reproduce. But now we're on the verge of being able to change the human species with genetic engineering. We will, at some point, try to enhance our intelligence by isolating the genes responsible for higher intelligence and greater analytical ability. And we'll spot those genes before we understand how they work—and long before we can correctly emulate them in digital programs. Artificial selection, instead of natural selection, will change our genetic makeup.

Our future is probably enhanced biological intelligence, not machine intelligence. And it's there that the dangers and/or benefits lie. We might, for example, select particular genes (or even create new genes) that we think will increase intelligence, while not really understanding how particular gene combinations work. Could we unknowingly begin a process that could alter the best human qualities? While striving for higher intelligence, could we somehow genetically diminish our capacity for compassion or our inherent need for social bonding? How might the human species be changed in the long run? The qualities that got us here—the curiosity, the intelligence, the compassion and cooperation resulting from our need for social bonding—involve a complex combination of genes. Could these be produced through artificial genetic selection? Could we lose them? Such worrying may not stop some scientists from deciding to use artificial selection. What will our grandchildren be like then?

REALLY GOOD HACKS

NEIL GERSHENFELD
Physicist; director, MIT's Center for Bits and Atoms; author, *FAB*

Something about discussion of artificial intelligence seems to displace human intelligence. The extremes of the arguments that AI is either our salvation or damnation are a sure sign of the impending irrelevance of this debate.

Disruptive technologies start as exponentials, which means the first doublings can appear inconsequential because the total numbers are small. Then comes a revolution when the exponential explodes, along with exaggerated claims and warnings to match, but it's a straight extrapolation of what's been apparent on a log plot. That's roughly when growth limits kick in, the exponential crosses over to a sigmoid, and the extreme hopes and fears disappear. That is what we're now living through with AI. The size of commonsense databases that can be searched, or the number of inference layers that can be trained, or the dimension of feature vectors that can be classified have all been making progress that seems discontinuous to someone who hasn't been following them.

Notably absent from either side of the debate are the people making many of the most important contributions to this progress. Advances like random matrix theory for compressed sensing, convex relaxations for heuristics for intractable problems, and kernel methods in high-dimensional function approximation are fundamentally changing our understanding of what it means to understand something.

The evaluation of AI has been an exercise in moving goalposts. Chess was conquered by analyzing more moves, *Jeopardy!* was won by storing more facts, natural-language translation was accomplished by accumulating more examples. These advances suggest that the secret of AI is likely to be that there isn't a secret. Like so much else in biology, intelligence appears to be a collection of really good hacks. There's a vanity in thinking that our consciousness is the defining attribute of our uniqueness as a species, but there's growing empirical evidence from studies of animal behavior and cognition that self-awareness evolved continuously and can be falsified in a number of other species. There's no reason to accept a mechanistic explanation for the rest of life while declaring one part of it to be off-limits.

We've long since become symbiotic with machines for thinking; my ability to do research rests on tools that help me to perceive, remember, reflect, and communicate. Asking whether or not they're intelligent is as fruitful as asking how I know I exist—amusing philosophically but not testable empirically.

Asking whether or not they're dangerous is prudent, as it is for any technology. From steam trains to gunpowder to nuclear power to biotechnology, we've never not been simultaneously doomed and about to be saved. In each case, salvation has lain in the much more interesting details rather than a simplistic yes/no argument for or against. We ignore the history of AI and everything else if we think this issue will be any different.

THE AIRBUS AND THE EAGLE

DANIEL L. EVERETT
Linguist; dean of arts and sciences, Bentley University; author, *Language: The Cultural Tool*

The more we learn about cognition, the stronger becomes the case for understanding human thinking as the nexus of several factors, as the emergent property of the interaction of the human body, human emotions, culture, and the specialized capacities of the entire brain. One of the greatest errors of Western philosophy was to buy into the Cartesian dualism of the famous statement, "I think, therefore I am." It's no less true to say, "I burn calories, therefore I am." Even better would be to say, "I have a human evolutionary history, therefore I can think about the fact that I am."

The mind is never more than a placeholder for things we don't understand about how we think. The more we use the solitary term *mind* to refer to human thinking, the more we underscore our lack of understanding. At least this is an emerging view of many researchers in fields as varied as neuroanthropology, emotions research, embodied cognition, radical embodied cognition, dual-inheritance theory, epigenetics, neurophilosophy, and the theory of culture.

For example, in the laboratory of Professor Martin Fischer at the University of Potsdam, interesting research is being done on the connection of the body and mathematical reasoning. Stephen Levinson's group at the Max Planck Institute for Psycholinguistics in Nijmegen has shown how culture can affect navigational abilities, a vital cognition function of most species.

In my own research, I'm looking at the influence of culture on the formation of what I refer to as "dark matter of the mind," a set of knowledges, orientations, biases, and patterns of thought that affect our cognition profoundly and pervasively.

If human cognition is indeed a property that emerges from the intersection of our physical, social, emotional, and data-processing abilities, then intelligence as we know it in humans is almost entirely unrelated to "intelligence" devoid of these properties.

I believe in artificial intelligence as long as we realize it's artificial. Comparing computation-problem solving, chess playing, reasoning, and so on to human thinking is like comparing the flight of an Airbus 320 to an eagle's. It's true that they both temporarily defy the pull of gravity, that they're both subject to the physics of the world in which they operate, and so on, but the similarities end there. Bird flight and airplane flight shouldn't be confused.

The reasons artificial intelligence isn't real intelligence are many. First, there's meaning. Some claim to have solved this problem, but they haven't, really. This "semantics problem" is, as John Searle pointed out years ago, why a computer running a translation program converting English into Mandarin speaks neither English nor Mandarin. No computer can learn a human language—only bits and combinatorics for special purposes. Second, there's the problem of what Searle calls the background and what I refer to as dark matter, or what some philosophers intend by the phrase *tacit knowledge*.

We learn to reason in a cultural context, whereby culture means a system of violable, ranked values, hierarchically structured knowledges, and social roles. We can do this not only because we have an amazing ability to perform what appears to be Bayesian inferencing across our experiences but also because of our emotions, our sensations, our proprioception, and our strong social ties. There's no computer with cousins and opinions about them.

Computers may be able to solve a lot of problems. But they cannot love. They cannot urinate. They cannot form social bonds because they're emotionally driven to do so. They have no romance. The popular idea that we may someday be able to upload our memories to the Internet and live forever is silly—we'd need to upload our bodies as well. The idea that comes up in discussions about artificial intelligence—that we should fear that machines will control us—is but a continuation of the idea of the religious "soul," cloaked in scientific jargon. It detracts from real understanding.

Of course, one ought never to say what science cannot do. Artificial intelligence may one day become less artificial by recreating bodies, emotions, social roles, values, and so on. But until it does, it will still be useful for vacuum cleaners, calculators, and cute little robots that talk in limited, trivial ways.

HUMANNESS

DOUGLAS COUPLAND
Writer, artist, designer; author, *Worst. Person. Ever.*

Let's quickly discuss larger mammals. Take dogs: We know what a dog is, and we understand "dogginess." Look at cats: We know what cats are and what "cattiness" is. Now take horses. Suddenly it gets harder. We know what a horse is, but what is "horsiness"? Even my friends with horses have trouble describing horsiness to me. And now take humans: What are we? What is "humanness"?

It's sort of strange, but here we are, 7 billion of us now, and nobody really knows the full answer to these questions. One undeniable thing we humans do, though, is make things, and through these things we find ways of expressing humanness— ways we didn't previously know of. The radio gave us Hitler and the Beach Boys. Barbed wire and air-conditioning gave us western North America. The Internet gave us a vanishing North American middle class and kitten GIFs.

It's said that new technologies alienate people, but the thing is, UFOs didn't land and hand us new technologies—we made them ourselves, and thus they can only ever be, well, humanating. And this is where we get to AI. People assume that AI, or machines that think, will have intelligence alien to our own, but that's not possible. In the absence of benevolent space aliens, only we humans will have created any nascent AI, and thus it can only mirror, in whatever manner, our humanness or specieshood. So when people express concern about alien intelligence, or the Singularity, what I think they're really expressing is angst about those unpretty parts of our collective being that currently remain unexpressed but will become, somehow, dreadfully apparent with AI.

As AI will be created by humans, its interface will be anthropocentric, just as AI designed by koala bears would be koalacentric. This means AI software will be humankind's greatest coding kludge, as we try molding it to our species' specific needs and data. Fortunately, anything smart enough to become sentient will probably be smart enough to rewrite itself from AI into cognitive simulation, at which point our new AI could become, for better or worse, even more human. We all hope for a Jeeves & Wooster relationship with our sentient machines, but we also need to prepare ourselves for a Manson & Fromme relationship; they're human too.

Personally I wonder whether the software needed for AI will be able to keep pace with the hardware in which it can live. Possibly the smart thing for us to do right now would be to set up a school whose sole goal is to imbue AI with personality, ethics, and compassion. It's certainly going to have enough data to work with, once it's born. But how best to deploy your grade-six report card, all of Banana Republic's returned merchandise data for 2037, and all of Google Books?

With the start of the Internet, we mostly had people communicating with other people. As time goes by, we increasingly have people communicating with machines. We all get excited about AI possibly finding patterns deep within metadata, and as the push grows to decode those profound volumes of metadata, the Internet will become largely about machines speaking with other machines—and what they'll be talking about, of course, is us, behind our backs.

MANIPULATORS AND MANIPULANDA

JOSH BONGARD

Associate professor of computer science, University of Vermont; coauthor (with Rolf Pfeifer), *How the Body Shapes the Way We Think*

Place a familiar object on a table in front of you, close your eyes, and manipulate that object so that it hangs upside down above the table. Your eyes are closed so that you can focus on your thinking. Which way did you reach out, grasp, and twist that object? What sensory feedback did you receive to know whether you were succeeding or failing? Now close your eyes again and think about manipulating someone you know into doing something he might not want to do. Again, observe your own thinking: What strategies might you employ? If you implement those strategies, how will you distinguish success from stalemate?

Although much recent progress has been made in building machines that sense patterns in data, most people feel that general intelligence involves *action*—reaching some desired goal, or failing that, keeping one's options open. It's hypothesized that this embodied approach to intelligence allows humans to use physical experiences (such as manipulating objects) as scaffolding for learning more subtle abilities (such as manipulating people). But our bodies shape the kinds of physical experiences we have. For example, we can manipulate only a few objects at once, because we have only two hands. Perhaps this limitation also constrains our social abilities in ways we have yet to discover. The cognitive linguist George Lakoff taught us that we can find clues to the body-centrism of thinking in metaphors: We counsel one another not to "look back" in anger because, based on our bias to walk in the direction of our forward-facing eyes, past events tend to literally be behind us.

So in order for machines to think, they must act. And in order to act they must have bodies to connect physical and abstract reasoning. But what if machines don't have bodies like ours? Consider Hans Moravec's hypothetical Bush Robot: Picture a shrub in which each branch is an arm and each twig is a finger. This robot's fractal nature would allow it to manipulate thousands or millions of objects simultaneously. How might such a robot differ in its thinking about manipulating people, compared with the way people think about manipulating people?

One of many notable deficiencies in human thinking is dichotomous reasoning—believing something is black or white rather than considering its particular shade of gray. But we're rigid and modular creatures; our branching set of bones houses fixed organs and supports fixed appendages with specific functions. What about machines that aren't so black and white? Thanks to advances in materials science and 3-D printing, soft robots are starting to appear. Such robots can change their shape in extreme ways; they may in the future be composed of 20 percent battery and 80 percent motor at one place on their surface, 30 percent sensor and 70 percent support structure at another, and 40 percent artificial material and 60 percent biological matter someplace else. Such machines may be much better able to appreciate gradations than we can.

Let's go deeper. Most of us have no problem using the singular pronoun *I* to refer to the tangle of neurons in our heads. We know exactly where we end and the world—and other people—begins. But consider modular robots, small cubes or spheres that can physically attach and detach to one another at will. How would such machines approach the self/nonself discrimination problem? Might such machines be able to empathize more strongly with other machines (and maybe even people) if they could physically attach to them or even become part of them?

That's how I think machines will think: in a familiar way because they'll use their bodies as tools to reason about the world, yet in an alien way because bodies different from human ones will lead to very different modes of thought.

But what do I think about thinking machines? I find the ethical side of thinking machines straightforward. Their dangerousness will correlate exactly with how much leeway we give them in fulfilling the goals we set for them. Machines told to "detect and pull broken widgets from the conveyer belt in the best way possible" will be quite useful, intellectually uninteresting, and likely to destroy more jobs than they create. Machines instructed to "educate this recently displaced worker"—or young person—"in the best way possible" will create jobs and possibly inspire the next generation. Machines commanded to "survive, reproduce, and improve in the best way possible" will give us the most insight into all the different ways in which entities may think, but they will probably give us humans a very short window of time in which to relish that insight. AI researchers and roboticists will sooner or later discover how to create all three of those species. Which ones we wish to call into being is up to us.

ARE WE THINKING MORE LIKE MACHINES?

ZIYAD MARAR
Global publishing director, SAGE; author,
Intimacy: Understanding the Subtle Power of Human Connection

There's something old-fashioned about visions of the future. The majority of predictions, like three-day weeks, personal jet packs, and the paperless office, tell us more about the times in which they were proposed than about contemporary experience. When people point to the future, we'd do well to run an eye back up the arm to see who's doing the pointing.

The possibility of artificial general intelligence has long invited such crystal-ball gazing, whether utopian or dystopian in tone. Yet speculations on this theme have reached such a pitch and intensity in the last few months alone (enough to trigger an *Edge* Question, no less) that this may reveal something about ourselves and our culture today.

We've known for some time that machines can outthink humans in a narrow sense. The question is whether they do so in any way that could or should ever resemble the baggier mode of human thought. Even when dealing with as tame a domain as chess, the computer and the human diverge widely.

"Tame" problems (like establishing the height of a mountain), which are well formulated and have clear solutions, are good grist for the mill of narrow, brute-force thinking. Sometimes even narrower thinking is called for, when huge data sets can be mined for correlations, leaving aside the distraction of thinking about underlying causes.

But many of the problems we face—from challenging inequality to choosing the right school for our children—are "wicked," in that they don't have right or wrong answers (though we hope they have better or worse ones). They're uniquely contextual and have complex overlapping causes that change based on the level of explanation used. Those problems don't suit narrow, computational thinking well. In blurring facts with values, they resemble the messy emotion-riddled thinking that reflects the human minds that conjured them up.

To tackle wicked problems requires peculiarly human judgment, even if these judgments are illogical in some sense—especially in the moral sphere. Notwithstanding Joshua Greene and Peter Singer's logical urging of a consequentialist frame of mind, one that a computer could reproduce, the human tendency to distinguish acts from omissions and to blur intentions with outcomes (as in the principle of double effect) means we need solutions that will satisfy the instincts of human judges if they're to be stable over time.

And that very feature of human thinking (shaped by evolutionary pressures) points to the widest gulf between machine and human thinking. Thinking is not motivated without preferences, and machines don't have those on their own. Only minds that comprehend cause and effect conjure up motives. So if goals, wants, values are features of human minds, then why predict that artificial superintelligences will become more than tools in the hands of those who program in those preferences?

If the welter of prognostications about AI and machine learning tells us anything, I don't think it's about how a machine will emulate a human mind anytime soon. We can do that easily enough just by having more children and educating them. Rather, it tells us that our appetites are shifting.

We're understandably awed by what sheer computation has achieved and will achieve; I'm happy to jump on the driverless virtual-reality bandwagon that careens off into that overpredicted future. But this awe is leading to a tilt in our culture. The digital republic of letters is yielding up engineering as the thinking metaphor of our time. In its wake lies the once complacent, now anxious figure with a more literary, less literal cast of mind. We're cleaning up our act, embarrassed by the fumbling inconclusiveness of messy thinking. It's unsurprising to hear that the United Kingdom's education secretary recently advised teenagers to steer away from arts and humanities in favor of STEM disciplines if they're to flourish. The sheer obviousness of a certain kind of progress has made narrow thinking gleam with a new and addictive luster.

But something's lost as whole fields of inquiry succeed or fail by the standard of narrow thinking, and a new impediment is created. Alongside the true, we need to think well about the good and the beautiful—and, indeed, the wicked. This requires vocabularies that better reflect our crooked timber (whether thought of, by turns, as bug or feature). Meanwhile, the understandable desire to upgrade those wicked problems to mere tame ones is leading us to tame ourselves.

JUST A NEW FRACTAL DETAIL IN THE BIG PICTURE

BRIAN ENO

Artist; composer; recording producer, U2, Coldplay, Talking Heads, Paul Simon; recording artist

Today I'm at my country cottage.

When the central heating takes effect, I'll get up and make myself some tea and porridge, to which I'll add some nuts and fruit. I'll switch on the *World Service* to hear the news, and then make a few phone calls about damp-proofing. And I'll probably plant the daffodil bulbs for spring (it says on the packet they should go in now). I think I'll then go to the supermarket and get some things for lunch and dinner, and perhaps take a bus into Norwich to look at getting a new bed. I don't have broadband in the cottage, so I'll also check my e-mails in Norwich, prebook a train back to London, and pay an electricity bill by electronic transfer.

And here's what I won't understand about all this. I won't understand how the oil that drives my central heating got from a distant oil field to my house. I won't know how it was refined into heating oil or what commercial transactions were involved. I won't know how the burner works. I won't know where my porridge or tea or nuts came from or how they got to me. I won't know how my phone works, or how my digital radio works, or how the news it relays to me was gathered or edited. I also won't understand the complexities of organizing a bus or train service, and I couldn't repair any of the vehicles involved. I won't really understand how a supermarket chain is run, or how beds are mass-produced, or how Wi-Fi works, or exactly what happens when I press "send" on my e-mail or transfer money electronically. And as for running an energy utility company, or putting in damp-proofing, or hybridizing daffodils to get these particular varieties, or why exactly I shouldn't plant them later than December—I won't understand any of that either.

Now here's the funny thing. I won't be in the least troubled by my vast ignorance about almost everything I'll be doing this morning. I'm used to it: I've been getting more and more ignorant all my life. I have a huge amount of experience in being ignorant and not worrying about it. In fact, what I call "understanding" turns out to be "managing my ignorance more effectively."

My untroubled attitude results from my almost absolute faith in the reliability of the vast supercomputer I'm permanently plugged into. It was built with the intelligence of thousands of generations of human minds, and they're still working at it now. All that human intelligence remains alive, in the form of the supercomputer of tools, theories, technologies, crafts, sciences, disciplines, customs, rituals, rules of thumb, arts, systems of belief, superstitions, work-arounds, and observations that we call Global Civilization.

Global Civilization is something we humans created, though none of us really know how. It's out of the individual control of any of us—a seething synergy of embodied intelligence that we're all plugged into. None of us understands more than a tiny sliver of it, but by and large we aren't paralyzed or terrorized by that fact—we still live in it and make use of it. We feed it problems—such as "I want some porridge"—and it miraculously offers us solutions that we don't really understand. What does that remind you of?

I read once that human brains began shrinking about 10,000 years ago and are now as much as 15 percent smaller than they were then. This corresponds with the point at which humans stopped having to be multicompetent individuals able to catch their own food and light their own fires and create their own tools and could instead become specialists—part of a larger community of humans who between them could do all the things that needed doing. Isn't the vast structure of competencies and potentialities thus created indistinguishable from "artificial intelligence"? The type that digital computers make is just a new fractal detail in the big picture, just the latest step. We've been living happily with artificial intelligence for thousands of years.

eGAIA, A DISTRIBUTED TECHNICAL-SOCIAL MENTAL SYSTEM

MARTI HEARST
Computer scientist, UC Berkeley, School of Information; author, *Search User Interfaces*

We will find ourselves in a world of omniscient instrumentation and automation long before a stand-alone sentient brain is built—if it ever is. Let's call this world *eGaia* for lack of a better word. In eGaia, electronic sensors (for images, sounds, smells, vibrations, all you can think of) are pervasive, and able to anticipate and arrange for the satisfaction of individuals' needs and allow for notification of all that's happening to those who need to know. Automation allows for cleaning of rooms and buildings, driving of vehicles, monitoring traffic, making and monitoring of goods, and even spying through windows (with tiny flying sensors). Already, major urban places are covered with visual sensors, and more monitoring is coming. In Copenhagen, LED-based streetlights will turn on only when they sense someone is biking down the road, and future applications of this network of sensors might include notifying when to salt the road or empty the trash, and, of course, alerting the authorities when suspicious behavior is detected on a street corner.

In eGaia, the medical advances will be astounding—synthetic biology makes smart machines that fix problems within our bodies; intelligent implants monitor and record current and past physical states. Brain-machine interfaces continue to be improved, initially for physically impaired people but eventually to provide a seamless boundary between people and the monitoring network. And virtual-reality-style interfaces will continue to become more realistic and immersive.

Why won't a stand-alone sentient brain come sooner? The amazing progress in spoken-language recognition—unthinkable ten years ago—derives in large part from having access to huge amounts of data and huge amounts of storage and fast networks. The improvements we see in natural-language processing are based on mimicking what people do, not understanding or even simulating it. It's not owing to breakthroughs in understanding human cognition or even significantly different algorithms. But eGaia is already partly here, at least in the developed world.

This distributed nerve-center network, an interplay among the minds of people and their monitoring electronics, will give rise to a distributed technical-social mental system the likes of which has not been experienced before.

THE HIVE MIND

CHRIS ANDERSON
Curator, TED conferences, TED talks

Thinking is our superpower. We're not the strongest, fastest, largest, or hardiest species. But we can model the future and act intentionally to realize the future we model. Somehow it's *this* power, not the ability to fly high, dive deep, roar loudly, or produce millions of babies, which has allowed its lucky recipients to visibly (as in literally visible from space) take over the planet. So if we succeed in building something that possesses our superpower, except more so, it will turn out to be a very big deal. Think about this question: In 1,000 years' time, will *Homo sapiens* plausibly be (a) the dominant intelligent force on Earth?, or (b) a historical footnote, the biological species that birthed intelligence?

Answer (a) seems unlikely to me. But if (b) is true, would that be a bad thing?

We all know how flawed humans are. How greedy, irrational, and limited in our ability to act collectively for the common good. We're in danger of wrecking the planet. Does anyone thoughtful really want humanity to be evolution's final word?

It all depends on how the transition goes. Power changes in many ways. There's violent suppression—what we presumably did to the Neanderthals. There are many scenarios where superintelligence takes us out just as unpleasantly.

But perhaps these scenarios ignore a key fact about intelligence. Intelligence doesn't reach its full power in small units.

Every additional connection and resource can help expand its power. A person can be smart, but a society can be smarter still. Your website is amazing, but Google connects that amazingness to a million other sites, and, lo and behold, all humanity's knowledge is there at your fingertips.

By that logic, intelligent machines of the future wouldn't destroy humans. Instead, they would tap into the unique contributions that humans make. The future would be one of ever richer intermingling of human and machine capabilities. I'll take that route. It's the best of those available.

Some of it will be glorious. And some uncomfortable. Maybe a few people won't appreciate being asked by some hybrid überintelligence to produce offspring genetically edited for higher creativity and less aggression, while enhanced by silicon implants. Or maybe the gorgeous 3-D simulation of their prospective offspring will convince them to proceed joyfully. Maybe people will look back nostalgically on the days when they used to own their time and could afford to page aimlessly through a pleasurable book just for the hell of it. But the astounding explosion of knowledge and imagination open to all will, most days, seem a fair substitute. One thing's for sure. Our own distinctive contribution to the ever more mind-boggling whole will gradually fade. And by that time we may not care.

It's already happening, by the way. I wake up in the morning, make my tea, and then drift over to my computer, which is calling to me. I flick it open and instantly I'm connected to 100 million other minds and machines around the world. I then spend forty-five minutes responding to its irresistible invitations. I initiate this process of my own free will. But then I surrender much of my will to the machine. So do you. Together we're semiunconsciously creating a hive mind of vastly greater power than this planet has ever seen—and vastly less power than it will soon see.

"Us versus the machines" is the wrong mental model. There's only one machine that really counts. Like it or not, we're all—us and our machines—becoming part of it: an immense connected brain. Once we had neurons. Now we're becoming the neurons.

THE GLOBAL ARTIFICIAL INTELLIGENCE IS HERE

ALEX (SANDY) PENTLAND

Toshiba Professor of Media Arts & Sciences, MIT; director, Human Dynamics Lab & Media Lab Entrepreneurship Program; author, *Social Physics: How Good Ideas Spread—The Lessons from a New Science*

The Global Artificial Intelligence (GAI) has already been born. Its eyes and ears are the digital devices all around us: credit cards, land-use satellites, cell phones, and of course the pecking of billions of people using the Web. Its central brain is rather like a worm at the moment—nodes that combine some sensors and some effectors—but the whole is far from what you'd call a coordinated intelligence.

Already many countries are using this infant nervous system to shape people's political behavior and "guide" the national consensus: China's Great Firewall, its siblings in Iran and Russia, and of course both major political parties in the United States. The national intelligence and defense agencies form a quieter, more hidden part of the GAI, but despite being quiet they're the parts that control the fangs and claws. More visibly, companies are beginning to use this newborn nervous system to shape consumer behavior and increase profits.

While the GAI is newborn, it has very old roots. The fundamental algorithms and programming of the emerging GAI have been created by the ancient guilds of law, politics, and religion. This is a natural evolution, because creating a law is just specifying an algorithm, and governance via bureaucrats is how you execute the program of law. Most recently, newcomers such as merchants, social crusaders, and even engineers have dared to add their flourishes to the GAI. The results of all these laws and programming are an improvement over Hammurabi, but we're still plagued by lack of inclusion, transparency, and accountability, along with poor mechanisms for decision making and information gathering.

However, in the last decades, the evolving GAI has begun to use digital technologies to replace human bureaucrats. Those with primitive programming and mathematical skills—namely lawyers, politicians, and many social scientists—have become fearful of losing their positions of power and so are making all sorts of noise about the dangers of allowing engineers and entrepreneurs to program the GAI. To my ears, the complaints of the traditional programmers sound rather hollow, given their repeated failures across thousands of years.

If we look at newer, digital parts of the GAI, we can see a pattern. Some new parts are saving humanity from the mistakes of the traditional programmers: Land-use space satellites alerted us to global warming, deforestation, and other environmental problems and gave us the facts to address those harms. Similarly, statistical analyses of health care, transportation, and work patterns have given us a worldwide network that can track global pandemics and guide public health efforts. On the other hand, some of the new parts—such as the Great Firewall, the NSA, and the U.S. political parties—are scary, because of the possibility that a small group of people can potentially control the thoughts and behavior of very large groups of people, perhaps without those people even knowing they're being manipulated.

What this suggests is that it isn't the Global Artificial Intelligence itself that's worrisome, it's how it's controlled. If the control is in the hands of just a few people, or if the GAI is independent of human participation, then the GAI can be the enabler of nightmares. But if control is in the hands of a large and diverse cross section of people, then the GAI's power is likely to be used to address problems faced by the entire human race. It's to our common advantage that the GAI becomes a distributed intelligence with a large and diverse set of humans providing guidance.

Creation of an effective GAI is critical, because today the human race faces many extremely serious problems. The GAI we've developed over the last 4,000 years, mostly made up of politicians and lawyers executing algorithms and programs developed centuries ago, is not only failing to address these serious problems but is threatening to extinguish us.

For humanity as a whole to first achieve and then sustain an honorable quality of life, we need to carefully guide the development of our GAI. Such a GAI might be in the form of a reengineered United Nations that uses new digital intelligence resources to enable sustainable development. But because existing multinational governance systems have failed so miserably, such an approach may require replacing most of today's bureaucracies with artificial intelligence "prosthetics"—digital systems that reliably gather accurate information and ensure that resources are distributed according to plan.

We already see this digital evolution improving the effectiveness of military and commercial systems, but it's interesting to note that as organizations use more digital prosthetics they also tend to evolve toward more distributed human leadership. Perhaps instead of elaborating traditional governance structures with digital prosthetics, we'll develop new, better types of digital democracy.

No matter how a new GAI develops, two things are clear. First, without an effective GAI, achieving an honorable quality of life for all of humanity seems unlikely. To vote against developing a GAI is to vote for a more violent, sick world. Second, the danger of a GAI comes from concentration of power. We must figure out how to build broadly democratic systems that include both humans and computer intelligences. It's critical to start building and testing GAIs that both solve humanity's existential problems and ensure equality of control and access. Otherwise we may be doomed to a future full of environmental disasters, wars, and needless suffering.

WILL COMPUTERS BECOME LIKE THINKING, TALKING DOGS?

RANDOLPH NESSE
Foundation Professor of Life Sciences, director,
Center for Evolution and Medicine, Arizona State University;
coauthor (with George C. Williams), *Why We Get Sick*

Thinking machines are evolving before our eyes. We want to know where they're headed. To find out, we need to look inward, since our desires are the forces that shape them. Alas, we can see ourselves only through a glass darkly. We didn't even anticipate that e-mail and social media would take over our lives. To see where thinking machines are headed, we need to look into the unforgiving mirror the Internet holds up to our nature.

Like the processed foods on grocery store shelves, Internet content is a product of selection for whatever sells. Every imaginable image, sound, and narrative gets posted, along with much that was previously unimaginable. The variations we ignore are selected out. Whatever grabs eyeballs is reposted with minor variations that evolve to whatever maximizes the duration of our attention.

That we can't tear ourselves away should be no surprise. Media content evolves to snare our attention, just as snacks and fast food evolve to become irresistible. Many lives are now as overstuffed with social media as they are with calories. We click and pop information bonbons into our minds the same way we pop chocolates into our mouths.

Enter thinking machines. They, too, are evolving. They will change faster and more radically when software is no longer designed, but instead evolves by selection among minor variations. However, until our brains coevolve with machines, our preferences will be the selection force. The machines that best satisfy them will evolve further, not to some Singularity but to become partners that fulfill our desires, for better or worse.

Many imagine coldly objective future computers, but no one likes a know-it-all. People will prefer modest, polite computers that are deeply subjective. Our machines won't contradict our inanities; they'll gently suggest, "That's an intriguing idea, but weren't you also thinking that . . ." Instead of objective sports stats, your machine will root with you for your team. If you get pulled over for speeding, your machine will blame the police and apologize for playing fast music. Machines that nag and brag will be supplanted by those expressing admiration for our abilities, even as they augment them. They'll encourage us warmly, share our opinions, and guide us to new insights so subtly that we'll imagine we thought of them ourselves.

Such relationships with machines will be different from those with real people, but they'll nonetheless be enduring and intense. Poets and pundits will spend decades comparing and contrasting real and virtual relationships, even while AIs increasingly become our trusted, treasured companions. Real people will find it hard to compete, but they'll have to. This will require behaving even more prosocially. The same process of social selection that has shaped human capacities for altruism and morality may become yet more intense, as people compete with machines to be interesting preferred partners. In fact, a look at living rooms where each family member is immersed in his or her own virtual world suggests that it's already hard to compete with machines.

In the short run, dogs stand the best of chance of competing with computers for our attention and affection. After several thousand years of selection, they're close to what we want them to be—loving, loyal, and eager to play and please. They're blissfully undistracted by their phones and tablets. Will computers evolve to become like thinking, talking dogs? We can hope. But I doubt that our machines will ever be furry and warm, with eyes that plead for a treat, a scratch, or a walk around the block. We'll prefer our dogs for a very long time. Our deepest satisfactions come, after all, not from what others do for us but from being appreciated for what we do for them.

THINKING MACHINES AND ENNUI

RICHARD E. NISBETT

Psychologist; Theodore M. Newcomb Distinguished University Professor, University of Michigan; author, *Mindware: Tools for Smart Thinking*

The first time I had occasion to think about what thinking machines might do to human existence was at a talk decades ago by a computer scientist at a Yale Psychology Department colloquium. The talk's topic was "What will it mean to humans' conception of themselves, and to their well-being, if computers are ever able to do everything better than humans can do—beat the greatest chess player, compose better symphonies than humans?"

"I want to make two things clear at the outset," the speaker said. "First, I don't know whether machines will ever be able to do those things. Second, I'm the only person in the room with the right to an opinion about that question." The latter statement was met with gasps and nervous laughter.

Decades later, it's no longer a matter of opinion that computers can do many of the astonishing things the speaker mentioned. And I'm worried that the answer to his question about what this will mean to us is that we'll feel utterly sidelined and demoralized by machines. I was sorry that Deep Blue beat Garry Kasparov at chess. I was momentarily depressed when Watson, its successor, defeated both its human *Jeopardy!* competitors. And of course we know that machines can already compose works that beat the socks off John Cage for interest and listenability!

We do have to worry about a devastating morale problem, when any work we might do can be done better by machines. What does it mean to airplane pilots that a machine can do their job better than they can? How long will it be before that occupation, like hundreds of others already, is made obsolete by machines? What will it mean to accountants, financial planners, and lawyers when machines can carry out, at the very least, nearly all of their bread-and-butter tasks more effectively and infinitely faster than they can? To physicians, physicists, and psychotherapists?

What will it mean when there's simply no meaningful work for any of us to do? When unsupervised machines plant and harvest crops? When machines can design better machines than any human could even think of? Or be a more entertaining conversationalist than even the cleverest of your friends?

Steve Jobs told us that it wasn't the customers' job to know what they want. Computers may be able to boast that it's not the job of humans to know what they want.

Like you, I love to read, listen to music, see movies and plays, experience nature. But I also love to work—to feel that what I do is fascinating, at least to me, and might possibly improve the lives of others. What would it mean to people like you and me if our work were simply pointless and there were only the other enjoyable things to do?

We already know what machine-induced obsolescence has meant to some of the world's peoples. It's no longer necessary for anyone to make their own bows and arrows and hunt animals, for any purpose other than recreation. Or to plant, cultivate, and harvest corn and beans. Some cultures built around such activities have collapsed and lost their meaning to the people who were shaped by them. Think, for example, of

Southwestern Indian tribes, or rural whites in South Dakota, Alabama, New Mexico, with their ennui, lassitude, and drug addiction. We have to wonder whether the world's people can face with equanimity the possibility of there being nothing to do other than to entertain themselves.

Which isn't to say that cultures couldn't evolve in some way as to make the absence of work acceptable, even highly satisfying. There are cultures in which there has been little to do in the way of work for eons, and people seem to have gotten along just fine. In some South Pacific cultures, people could get by with little other than waiting for a coconut to drop or wading into a lagoon to catch a fish. In some West African cultures, men didn't do anything you'd be likely to classify as work except for a couple of weeks a year, when they were essential for the planting of crops. And then there were the idle rich of, say, early twentieth-century England, with their endless rounds of card playing, the putting on of different costumes for breakfast, lunch, and dinner, and serial infidelities with really rather attractive people. Judging from PBS fare, that was pretty enjoyable.

So maybe the most optimistic possibility is that we're headed toward evolving cultures that will enable us to enjoy perpetual entertainment with no meaningful, productive work to do. However repellent that may seem to us, we have to imagine—hope, even—that it may seem a delightful existence to our great-great-grandchildren, who will pity us for our cramped and boring lives. Some would say the vanguard is already here: Portland, Oregon, has been described as the place where young people go to retire.

NACHES FROM OUR MACHINES

SAMUEL ARBESMAN
Complex systems scientist; senior scholar, Ewing Marion Kauffman Foundation; associate, Institute for Quantitative Social Science, Harvard University; author, *The Half-Life of Facts*

When I think about machines that think, while I'm interested in how they might become possible, I'm more interested in how, as a society, we might respond to them. For example, if they fail to exhibit anything we might take for self-awareness or sentience, then they're certainly clever but humanity is still on the cognitive pedestal.

But what about when these thinking machines are as smart as we are, or even far more intelligent? What if they're intelligent in ways foreign to our own patterns of thought? This isn't so unlikely, as computers are already very good at things we're not good at: They have better short- and long-term memories, they're faster at calculations, and they're not bound by the irrationalities that hamstring our minds. Extrapolate this out and we can see that thinking machines might be both smart and alien.

So how shall we respond? One response is to mark them as monsters—unspeakable horrors that can examine the unknown in ways we cannot. Many people might respond this way if we birth machines that think about the world in ways wildly foreign from ours.

But it needn't be so. I prefer a more optimistic response, that of *naches*, a Yiddish term meaning joy and pride, often vicarious pride: You have naches—or, as in Yiddish, you *shep naches*—

when your children graduate from college, or get married, or pass any similar milestone. These aren't your own accomplishments but you still have a great deal of pride and joy in them.

The same can be true with our machines. We might not understand their thoughts or discoveries or technological advances, but they're *our* machines, and their creators can shep naches from the accomplishments of such offspring—say, computer programs that generate sophisticated artworks or musical compositions. I imagine the programmers of this software are proud of the resulting piece of art or music even if they can't generate it themselves.

We can broaden this sense of naches. Many of us are sports fans and take pride in our team's wins even though we had nothing to do with them. Or we're excited when a citizen of our country takes the gold in the Olympics, or makes a new discovery and is awarded a prestigious prize. So, too, should it be with our thinking machines for all of humanity: We can root for what humans have created even though it wasn't our own personal achievement—and even if we can't fully understand it. Many of us are grateful for technological advances, from the iPhone to the Internet, even though we don't fully know how they work.

When our children do something amazing, something we can't really understand, we don't despair, or worry; we're delighted and grateful for their success. In fact, gratitude is how many of us respond to technology currently. We can't completely understand our machines, but they work in powerful and useful ways and we're grateful for this. We can respond similarly to our future technological creations, these thinking machines we might not fully understand. Rather than fear or worry, we should have naches from them.

NO SHARED THEORY OF MIND

GERALD SMALLBERG
Neurologist, Lenox Hill Hospital, New York City; playwright, Off-Off Broadway productions, *Charter Members* and *The Gold Ring*

My thinking about this year's question is tempered by the observation made by Mark Twain in *A Connecticut Yankee in King Arthur's Court:* "A genuine expert can always foretell a thing that is 500 years away easier than he can a thing that's only 500 seconds off." Twain was being generous: Forget the 500 seconds, we'll never know with certainty even one second into the future. However, humans can try contemplating the future, which provided *Homo sapiens* with its great evolutionary advantage. This talent for imagining a future has been the engine of progress, the source of creativity.

We've built machines that in simplistic ways are already "thinking" by solving problems or performing tasks we've designed. Currently they're subject to algorithms that follow rules of logic, whether it be "crisp" or "fuzzy." This intelligence, despite its vast memory and increasingly advanced processing mechanisms, is still primitive. In theory, as these machines become more sophisticated, they'll at some point attain a form of consciousness, defined for the purpose of this discussion as the ability to be aware of being aware—most likely by combining the properties of silicon and carbon with digital and analog parallel processing and possibly even quantum computing, and with networks that incorporate time delay.

This form of consciousness, however, will be devoid of subjective feelings or emotions. There are those who argue that feelings are triggered by the thoughts and images that have become paired with a particular emotion: Fear, joy, sadness, anger, and lust are examples of emotions; feelings include contentment, anxiety, happiness, bitterness, love, and hatred.

My opinion that machines will lack this aspect of consciousness is based on two considerations. The first is appreciating how we arrived at the ability to feel and have emotions. As human beings, we're the end product of evolution by natural selection—a process that arose in primitive organisms approximately 3.5 billion years ago. Over this vast eon of time, we're not unique in the animal kingdom in experiencing feelings and emotions. But in the last 150,000 to 300,000 years, *Homo sapiens* is singular in having evolved the ability to use language and symbolic thought as part of how we reason, in order to make sense of our experiences and view the world we inhabit. Feeling, emotion, and intellectual comprehension are inexorably intertwined with how we think. Not only are we aware of being aware, but also our ability to think enables us to remember a past and imagine a future. Using our emotions, feelings, and reasoned thoughts, we can form a Theory of Mind so that we can understand the thinking of other people, which in turn has enabled us to share knowledge as we have created societies, cultures, and civilizations.

The second consideration is that machines aren't organisms, and no matter how complex and sophisticated they become, they won't have evolved by natural selection. Regardless of how they're designed and programmed, their possession of feelings and emotions would be counterproductive to what will make them most valuable to us.

The driving force for building advanced intelligent machines will be the need to process and analyze incomprehensible amounts of future information and data to help us ascertain what's likely to be true from what's false, what's relevant from what's irrelevant. They will make predictions, since they, too, will be able to peer into the future while waiting (as will always be the case) for its cards to be revealed. They'll have to be totally rational agents in order to perform these tasks accurately and reliably.

In their decision analysis, a system of moral standards will be necessary. Perhaps it will be a calculus incorporating such utilitarian principles as the "The greatest happiness of the greatest number is the measure of right and wrong" along with the Golden Rule, the foundational precept that underlies many religions ("Treat others as one would like others to treat oneself"). The subjective values introduced by feelings and emotions would amount to a self-defeating strategy for solving the complex problems we'll continue to face as we weigh what's best for our own species along with the rest of life we share our planet with.

My experience as a clinical neurologist leads me to believe that we'll be unable to read machines' thoughts. But also they'll be incapable of reading ours. There will be no shared Theory of Mind. I suspect the closest we can come to knowing this most complex of states is indirectly, by studying the behavior of these superintelligent machines. They will have crossed that threshold when they start replicating and looking for an energy source solely under their control. If this should occur, and if I'm still around (a highly unlikely expectation), my judgment about whether it presages a utopian or dystopian future will be based on my thinking—biased as always, since it will remain a product of analytical reasoning colored by my feelings and emotions.

BLIND TO THE CORE OF HUMAN EXPERIENCE

ELDAR SHAFIR
William Stewart Tod Professor of Psychology and Public Affairs, Princeton University; coauthor (with Sendil Mullainathan), *Scarcity: Why Having Too Little Means So Much*

Thinking comes in many forms, from solving optimization problems and playing chess to having a smart conversation or composing a fine piece of music. But when I think about machines that purportedly think, I wonder about what they might be thinking when the topics are inherently human, as so many topics inherently are.

Consider Bertrand Russell's touching description in *What I Have Lived For*:

> Three passions, simple but overwhelmingly strong, have governed my life: the longing for love, the search for knowledge, and unbearable pity for the suffering of mankind. These passions, like great winds, have blown me hither and thither, in a wayward course, over a deep ocean of anguish, reaching to the very verge of despair.

Although Russell was a celebrated thinker, what he describes is, in one form or another, familiar to us all. But what would a machine make of this? Could it really feel a "longing for love" or an "unbearable pity for the suffering of mankind"? Could it be "blown . . . hither and thither . . . over a deep ocean of anguish, reaching to the very verge of despair"?

If we accept some version of the computer metaphor of the mind (and I do), then all these sentiments must ultimately be the products of physical processes, which, in theory, can be instantiated by a machine. But the topics themselves so often are human. If we agree that it's hard for males to fully understand maternal love; that the satiated may be unable to grasp what it feels like to starve; that the free may not fully comprehend what it's like to be imprisoned—well, then machines, no matter how well they "think," may be unable to think of many things. And those things are at the core of human experience. At the opera, we feel for Aida, who is horrified to hear herself call out "Ritorna vincitor," finding herself torn between her love for Radames and her devotion to her father and her people. Could a machine feel torn like Aida, or even moved like the rest of us when we see her beg the gods to pity her suffering? Can a machine experience fear of death, without living? Lust, without having sexual organs? Or the thoughts that come with headaches, wrinkles, or the common cold? It's easy to imagine a machine dressed in a Nazi uniform and another machine we can call "Sophie." But when the former forces the latter to make a horrific choice, can the first experience the sadism and the second an irreparable desperation of the kind rendered so palpable in Styron's story?

If machines cannot truly experience the sort of thinking that incorporates the passions and sorrows of the likes of Russell, or Aida, or Sophie; if they cannot experience the yearnings, desires, determination, and disgrace underlying the thinking of Nabakov's Humbert, Conrad's Kurtz, Melville's Ahab, or Tolstoy's Anna; if they cannot do any of that, then perhaps they cannot really fully think.

AN INTUITIVE THEORY OF MACHINE

CHRISTOPHER CHABRIS
Associate professor of psychology, Union College;
coauthor (with Daniel Simons), *The Invisible Gorilla:
How Our Intuitions Deceive Us*

I've often wondered why we human beings have so much trouble thinking straight about machines that think.

In the arts and entertainment, machines that can think are often depicted as simulacra of humans, sometimes down to the shape of the body and its parts, and their behavior suggests that their thoughts are much like our own. But thinking doesn't have to follow human rules or patterns to count as thinking. Examples of this abound: Chess computers outthink humans not because they think like humans think about chess except better, but because they think in an entirely different way. Useful language translation can be done without deep knowledge of grammar.

Evolution has endowed human beings with the ability to represent and reason about the contents of other human minds. By the time children start school, they can keep track of what different people know about the same set of facts (this is a prerequisite for lying). Later, as adults, we use this capacity to figure out how to negotiate, collaborate, and solve problems for the benefit of ourselves and others. This piece of mental equipment is often called Theory of Mind, and springs into action even in situations where there are no "minds" to represent. Videos of two-dimensional shapes moving around on computer screens can tell stories of love, betrayal, hate, and violence that exist entirely in the mind of the viewer, who temporarily forgets that polygons don't have emotions.

Maybe we have trouble thinking about thinking machines because we don't have a correspondingly intuitive Theory of Machine. Mentally simulating a simple mechanical device consisting of a few interlocking gears—say, figuring out whether turning the first gear will cause the last gear to rotate left or right, faster or slower—is devilishly difficult. Complex machines consisting of abstract algorithms and data are just as alien to our built-in mental faculties.

Perhaps this is why, when faced with the notion of thinking machines, we fall back on understanding them as though they were thinking beings—in other words, as though they were humans. We apply the best tools our mind has—namely, Theory of Mind and general-purpose reasoning. Unfortunately, the former isn't designed for this job and the latter is hampered by our limited capacities for attention and working memory. Sure, we have disciplines like physics, engineering, and computer science that teach us how to understand and build machines, including machines that think, but years of formal education are required to appreciate the basics.

A Theory of Machine module would ignore intentionality and emotion and instead specialize in representing the interactions of different subsystems, inputs, and outputs to predict what machines would do in different circumstances, much as Theory of Mind helps us to predict how other humans will behave.

If we did have Theory of Machine capacities built into our brains, things might be different. Instead, we seem condemned to see the complex reality of thinking machines, which think based on different principles from the ones we're used to, through the simplifying lens of assuming they'll be like thinking minds, perhaps reduced or amplified in capacity but essentially the same. Since we'll be interacting with thinking machines more and more as time goes on, we need to figure out how to develop better intuitions about how they work. Crafting a new module isn't easy, but our brains did it—by reusing existing faculties in a clever new way—when written language was invented. Perhaps our descendants will learn the skill of understanding machines in childhood as easily as we learned to read.

THINKING SALTMARSHES

URSULA MARTIN

Professor of computer science, Oxford University

Hiking toward the saltmarsh at dusk, I pause, confused, as the footpath seems to disappear into a long stretch of shallow muddy water, shining as it reflects the light of the setting sun. Then I notice a line of stepping-stones, visible only because their rough texture just ruffles the bright smooth surface of the water. And I set my pace to the rhythm of the stones, and walk on across the marsh to the sand dunes beyond.

Reading the watery marshland is a conversation with the past, with people I know nothing about, except that they laid the stones that shape my stride, and probably shared my dislike of wet feet.

Beyond the dunes, wide sands stretch across a bay to a village beyond. The receding tide has created strangely regular repeating patterns of water and sand, which echo a line of ancient wooden posts. A few hundred years ago salmon were abundant here, and the posts supported nets to catch them. A stone church tower provides a landmark, and I stride out cross the sands toward it to reach the village, disturbing noisy groups of seabirds.

The water, stepping-stones, posts, and church tower are the texts of a slow conversation across the ages. Path makers, salmon fishers, and even solitary walkers mark the land; the weather and tides, rocks and sand and water, creatures and plants respond to those marks; and future generations in turn respond to and change what they find.

Where then are the thinking machines? One can discuss the considerable challenges to artificial intelligence posed by scene analysis and route-finding across liquid marshes and shifting beaches; or in grasping narratives of the past set out not in neat parsable text but through worn stepping-stones and rotting wooden posts.

One can picture and debate a thinking machine to augment the experience of our solitary walker. Perhaps a cute robot companion splashing through the marsh and running out along the sand chasing the seabirds. Or a walker guided along the path by a thinking machine that integrates a buzz of data streams on paths, weather, and wildlife, to provide a cocoon of step-by-step instructions, nature notes, historical factoids, and fitness data, alongside alerts about privacy risks and the dangers of the incoming tide. Or a thinking machine that works out where the birds go in the summertime, or how to make the salmon abundant again.

But what kind of thinking machine might find its own place in slow conversations over the centuries, mediated by land and water? What qualities would such a machine need to have? Or what if the thinking machine was not replacing any individual entity but was used as a concept to help understand the combination of human, natural, and technological activities that create the sea's margin, and our response to it? The term *social machine* is currently used to describe endeavors that are purposeful interactions of people and machines—Wikipedia and the like—so the "landscape machine," perhaps.

Ah yes, purposeful. The purpose of the solitary walker may be straightforward—to catch fish, to understand birds, or merely to get home safely before the tide comes in. But what if the purpose of the solitary walker is no more than a solitary walk—to find balance, to be at one with nature, to enrich the imagination, or to feed the soul. Now the walk becomes a conversation with the past, not directly through rocks and posts and water but through words, through the poetry of those who have experienced humanity through rocks and posts and water and found the words to pass that experience on. So the purpose of the solitary walker is to reinforce those very qualities that make the solitary walker a human being, in a shared humanity with other human beings. A challenge indeed for a thinking machine.

KILLER THINKING MACHINES KEEP OUR CONSCIENCE CLEAN

KURT GRAY
Assistant professor of psychology, University of North Carolina, Chapel Hill

Machines have long helped us kill. From catapults to cruise missiles, mechanical systems have allowed humans to better destroy one another. Despite the increased sophistication of killing machines, one thing has remained constant—human minds are always morally accountable for their operation. Guns and bombs are inherently mindless, and so blame slips past them to the person who pulled the trigger.

But what if machines had enough of a mind that they could choose to kill on their own? Such a thinking machine could retain the blame for itself, keeping clean the consciences of those who benefit from its work of destruction. Thinking machines may improve the world in many ways, but they may also let people get away with murder.

Humans have long sought to distance themselves from acts of violence, reaping the benefits of harm without sullying themselves. Machines not only increase destructive power but also physically obscure our harmful actions. Punching, stabbing, and choking have been replaced by the more distant—and tasteful—actions of button pressing or lever pulling. However, even with the increased physical distance allowed by machine intermediaries, our minds continue to ascribe blame to those people behind them.

Studies in moral psychology reveal that humans have a deep-seated urge to blame someone or something in the face of suffering. When others are harmed, we search not only for a cause but a mental cause—a thinking being who chose to cause the suffering. This thinking being is typically human but need not be. In the aftermath of hurricanes and tsunamis, people often blame the hand of God, and in some historical cases people have even blamed livestock—French peasants once placed a pig on trial for murdering a baby.

Generally, our thirst for blame requires only a single thinking being. When we find one thinking being to blame, we're less motivated to blame another. If a human is to blame, there's no need to curse God. If a low-level employee is to blame, there's no need to fire the CEO. And if a thinking machine is to blame for someone's death, then there's no need to punish the humans who benefit.

Of course, for a machine to absorb blame, it must be a legitimate thinker and act in new, unpredicted ways. Perhaps machines could never do something truly new, but the same argument applies to humans "programmed" by evolution and their cultural context. Consider children, who are undoubtedly programmed by their parents and yet—through learning—are able to develop novel behavior and moral responsibility. Like children, modern machines are adept at learning, and it seems inevitable that they'll develop contingencies unpredicted by their programmers. Already, algorithms have discovered new things unguessed by the humans who created them.

Thinking machines might make their own decisions but shield humans from blame only when they decide to kill, standing between our minds and the destruction we desire.

Robots already play a large role in modern combat: Drones have killed thousands in the past few years but are currently fully controlled by human pilots. To deflect blame in this case, the drones must be governed by other intelligent machines; machines must learn to fly Predators all on their own.

This scenario may send shivers down spines (including mine), but it makes cold sense from the perspective of policy makers. If collateral damage can be blamed on the decisions of machines, then military mistakes are less likely to dampen someone's election chances. Moreover, if minded machines can be overhauled or removed—machine "punishment"— people will feel less need to punish those in charge, whether for fatalities of war, botched (robotic) surgeries, or (autonomous) car accidents.

Thinking machines are complex, but the human urge to blame is relatively simple. Death and destruction compel us to find a single mind to hold responsible. Sufficiently smart machines—if placed between destruction and ourselves—should absorb the weight of wrongdoing, shielding our own minds from others' condemnation. We should all hope that this prediction never comes true, but when advancing technology collides with modern understandings of moral psychology, dark potentials emerge. To keep clean our consciences, we need only to create a thinking machine and then vilify it.

WHEN THINKING MACHINES BREAK THE LAW

BRUCE SCHNEIER
Security technologist; fellow, Berkman Center for Internet and Society, Harvard Law School; chief technical officer, Co3 Systems, Inc.; author, *Data and Goliath: The Hidden Battles to Collect Your Data and Control Your World*

Last year, two Swiss artists programmed a Random Botnot Shopper, which every week would spend $100 in bitcoin to buy a random item from an anonymous Internet black market—all for an art project on display in Switzerland. It was a clever concept, except there was a problem. Most of the stuff the bot purchased was benign—fake Diesel jeans, a baseball cap with a hidden camera, a stash can, a pair of Nike trainers—but it also purchased ten ecstasy tablets and a fake Hungarian passport.

What do we do when a machine breaks the law? Traditionally, we hold the person controlling the machine responsible. People commit the crimes; the guns, lock picks, or computer viruses are merely their tools. But as machines become more autonomous, the link between machine and controller becomes more tenuous.

Who's responsible if an autonomous military drone accidentally kills a crowd of civilians? Is it the military officer who keyed in the mission, the programmers of the enemy detection software that misidentified the people, or the programmers of the software that made the actual kill decision? What if those programmers had no idea that their software was being used for military purposes? And what if the drone could improve its algorithms by modifying its own software based on what the entire fleet of drones learned on earlier missions?

Maybe our courts can decide where the culpability lies, but that's only because whereas current drones are autonomous, they're not very smart. As drones get smarter, their links to the humans who built them become more tenuous.

What if there are no programmers, and the drones program themselves? What if they're smart and autonomous and make strategic as well as tactical decisions on targets? What if one of the drones decides, based on whatever means it has at its disposal, that it will no longer maintain allegiance to the country that built it, and goes rogue?

Our society has many approaches, using both informal social rules and more formal laws, for dealing with people who won't follow the rules. We have informal mechanisms for small infractions and a complex legal system for larger ones. If you're obnoxious at my party, I won't invite you back. Do it regularly, and you'll be shamed and ostracized from the group. If you steal some of my stuff, I might report you to the police. Steal from a bank, and you'll almost certainly go to jail for a long time. A lot of this might seem more ad hoc than situation-specific, but we humans have spent millennia working this all out. Security is both political and social, but it's also psychological. Door locks, for example, work only because our social and legal prohibitions on theft keep the overwhelming majority of us honest. That's how we live peacefully together on a scale unimaginable for any other species on the planet.

How does any of this work when the perpetrator is a machine with whatever passes for free will? Machines probably won't have any concept of shame or praise. They won't refrain from doing something because of what other machines might think. They won't follow laws simply because it's the right thing to do, nor will they have a natural deference to authority. When they're caught stealing, how can they be punished? What does it mean to fine a machine? Does it make any sense at all to incarcerate it? And unless they're deliberately programmed with a self-preservation function, threatening them with execution will have no meaningful effect.

We're already talking about programming morality into thinking machines, and we can imagine programming other human tendencies into them, but we're certainly going to get it wrong. No matter how much we try to avoid it, we'll have machines that break the law.

This, in turn, will break our legal system. Fundamentally, our legal system doesn't prevent crime. Its effectiveness is based on arresting and convicting criminals after the fact—and on their punishment providing a deterrent to others. This fails if there's no punishment that makes sense.

We experienced an example of this after 9/11, when most of us first started thinking about suicide terrorists and the fact that ex post facto security was irrelevant to them. That was just one change in motivation, and look at how those actions affected the way we view security. Our laws will have the same problem with thinking machines, along with related problems we can't yet imagine. The social and legal systems that have dealt so effectively with human rule breakers will fail in unexpected ways in the face of thinking machines.

A machine that thinks won't always think in the ways we want it to. And we're not ready for the ramifications of that.

ELECTRIC BRAINS

REBECCA MACKINNON
Director, Ranking Digital Rights Project,
New America Foundation; cofounder, Global Voices;
author, *Consent of the Networked*

The Chinese word for *computer* translates literally as "electric brain." How do electric brains "think" today? As individual machines, still primitively by human standards. Powerfully enough in the collective. Networked devices and all sorts of things with electric brains embedded in them increasingly communicate with one another, share information, reach mutual "understandings," and make decisions. It's already possible for a sequence of data retrieval, analysis, and decision making, distributed across a cloud of machines in various locations, to trigger action by a single machine or set of machines in one physical place, thereby affecting (or in service of) a given human or group of humans.

Perhaps individual machines may never think in a way that resembles individual human consciousness as we understand it. But maybe someday large globally distributed networks of nonhuman things may achieve some sort of pseudo–Jungian collective consciousness. More likely, the collective consciousness of human networks and societies will be enhanced by—and increasingly intertwined with—a different sort of collective consciousness generated by networks of electric brains.

Will this be a good thing or a bad thing?

Both. Neither. Like the Internet we all use today, it depends whether you think human nature is fundamentally good or bad or both. The Internet doesn't transform or improve human nature. It magnifies, telescopes, enhances, empowers, and concentrates many aspects of human nature—from the altruistic and charitable to the criminal and evil. Get ready to add another dimension to what the Internet already does.

We already have what computer scientists like to call "attribution problems": identifying who's truly responsible for something that happens on or through the Internet (say, for example, a cyberattack on a government facility or multinational corporation). Those problems and debates are going to get even tougher very quickly.

We'll continue to ask many of the questions about human-rights implications of a much smarter and empowered cloud that we're asking today about the Internet and networked devices. Who gets to shape the technology we increasingly depend on for our economic, social, political, and religious lives? Who's responsible when somebody's rights are violated via those technologies, platforms, and networks? Who gets to hold whom accountable for violations including censorship, surveillance, incitement to physical violence, data-driven discrimination, etc.?

New questions: Will rival networks of thinking things, created by and connected closely to (note I don't say "controlled by") rival cultures, commercial alliances, religions, or polities, block connections between one another? Might they fight one another? How will artistic creation work? How will politics work? How will war work? Can censorship and surveillance be delegated to nonhuman networks so that humans can avoid taking responsibility for such things? (How convenient for our government and business leaders.) Can thinking networks of things instead be engineered so that direct human involvement or sign-off is required for certain types of actions taken by machines?

Will empowered smart clouds exacerbate global inequalities? Might they exacerbate global ideological and religious conflicts if we don't act to stop that? If we want to prevent the global digital divide from deepening, what early steps must be taken?

Will these networks be open or closed? Will any innovator from anywhere be able to plug something new into a network and communicate—or shall we say participate—without needing permission? Or will it be a controlled system, with certain companies or governments deciding who and what is allowed to connect at what price? Or will some systems be open and some closed?

Will smarter and more empowered global networks erode the power and legitimacy of nation-states beyond what the Internet has already done? Or might they extend the power of nation-states in new ways? Or enable the nation-state to evolve and ultimately survive in a digitally networked world?

We cannot assume good or humane outcomes just because the people who invent the technology or set the process in motion seem like well-intentioned, freedom-and-democracy-loving people. Such assumptions didn't work well for the Internet, and they won't work any better for whatever comes next.

ROBODOCTORS

GERD GIGERENZER

Psychologist; director, Center for Adaptive Behavior and Cognition, Max Planck Institute for Human Development, Berlin; author, *Risk Savvy: How to Make Good Decisions*

It's time for your annual checkup. Entering your doctor's office, you shake her cold hand, the metal hand of a machine. You're face-to-face with an RD, a certified robodoctor. Would you like that? No way, you might say. I want a real doctor, someone who listens to me, talks to me, and feels like me. A human being whom I can trust, blindly.

But think for a moment. In fee-for-service health care, a primary care physician may spend no more than five minutes with you. And during this short time, astonishingly little thinking takes place. Many doctors complain to me about their anxious, uninformed, noncompliant patients with unhealthy lifestyles who demand drugs advertised by celebrities on television and, if something goes wrong, threaten to turn into plaintiffs.

But lack of thinking doesn't simply affect patients: Studies consistently show that most doctors don't understand health statistics and thus cannot critically evaluate a medical article in their own field. This collective lack of thinking has its toll. Ten million U.S. women have had unnecessary Pap smears to screen for cervical cancer—unnecessary because they'd had a full hysterectomy and thus no cervix anymore. Every year, 1 million U.S. children have unnecessary CT scans, which expose them to radiation levels that cause cancer in some of them later in life. And many doctors ask men to undergo regular PSA screening for prostate cancer, despite the fact that virtually all medical organizations recommend against it because it has no proven benefit but can cause severe harm. Scores of men end up incontinent and impotent from subsequent surgery or radiation. All this adds up to a huge waste of doctors' time and patients' money.

So why don't doctors always recommend what's best for the patient? There are three reasons. First, as noted, some 70 to 80 percent of physicians don't understand health statistics. The cause? Medical schools across the world fail to teach statistical thinking. Second, in fee-for-service systems, doctors have conflicts of interest: They lose money if they don't recommend tests and treatments, even if these are unnecessary or harmful. Third, more than 90 percent of U.S. doctors admit to practicing defensive medicine—that is, recommending unnecessary tests and treatments that they wouldn't recommend to their own family members. They do this to protect themselves against you, the patient, who might pursue litigation. Thus, a doctor's office is packed with psychology that gets in the way of good care—self-defense, innumeracy, and conflicting interests. This threefold malady is known as the SIC syndrome. It undermines patient safety.

Does it matter? Based on data from 1984 and 1992, the Institute of Medicine estimated that some 44,000 to 98,000 patients die from preventable and documented medical errors every year in U.S. hospitals. Based on more recent data, from 2008 to 2011, Patient Safety America has updated this death toll to more than 400,000 per year. Nonlethal serious harm caused by these preventable errors occurs in an estimated 4 to 8 million Americans every year. The harm caused in private practice is not known. If fewer and fewer doctors have less and less time for patients and patient safety, this epidemic of harm will continue to spread. Ebola pales compared to it.

A revolution in health care is wanted. Medical schools should teach students the basics of health statistics. Legal systems should no longer punish doctors if they rely on evidence rather on convention. We also need incentive systems that don't force doctors to choose between making a profit and providing the best care for the patient. But this revolution hasn't happened, and there are few signs that it will.

So why not resort to a radical solution: robodoctors who understand health statistics, have no conflicts of interest, and aren't afraid of being sued (after all, they don't have to repay medical school debts and have no bank accounts to protect from litigation)? Let's go back to your annual checkup. You might ask the RD whether checkups reduce mortality from cancer, from heart disease, or from any other cause. Without hedging, the RD would inform you that a review of all existing medical studies shows that the answer is no, on all three counts. You might not want to hear that, because you're proud of conscientiously going for routine checkups after hearing the opposite from your human doctor, who may have had no time to keep up with medical science. And your RD would not order unnecessary CTs for your child, or Pap smears if you're a woman without a cervix, or recommend routine PSA tests without explaining the pros and cons if you're a man. Moreover, they can talk to multiple patients simultaneously and thus give you as much time as you need. Waiting time will be short and nobody will rush you out the door.

When we imagine thinking machines, we tend to think about better technology—about devices for self-monitoring blood pressure, cholesterol, or heart rate. My point is different. The RD revolution is less about better technology than about better psychology. That is, it entails thinking more about what's best for the patient and striving for best care instead of best revenues.

OK. Your next objection is that for-profit clinics will easily undercut this vision of propatient robots and program RDs so that they maximize profit rather than your health. You've put your finger on the essence of our health-care malady. But there's a psychological factor that will likely help. Patients often don't ask questions in consultations with human MDs because they rely on the dictum "Trust your doctor." But that rule doesn't necessarily apply to machines. After shaking an RD's icy hand, patients may well begin to think for themselves. Making people think is the best that a machine can achieve.

CAN MACHINES EVER BE AS SMART AS THREE-YEAR-OLDS?

ALISON GOPNIK
Psychologist, UC Berkeley; author, *The Philosophical Baby*

They may outwit Kasparov, but can machines ever be as smart as a three-year-old?

Learning has been at the center of the new revival of AI. But the best learners in the universe, by far, are still human children. In the last ten years, developmental cognitive scientists, often collaborating with computer scientists, have been trying to figure out how children could possibly learn so much so quickly.

One of the fascinating things about the search for AI is that it's been so hard to predict which parts would be easy or hard. At first, we thought that the quintessential preoccupations of the officially smart few, like playing chess or proving theorems—the corridas of nerd machismo—would prove hardest for computers. In fact they turn out to be easy. Things every dummy can do, like recognizing objects or picking them up, are much harder. And it turns out to be much easier to simulate the reasoning of a highly trained adult expert than to mimic the ordinary learning of every baby. So where are machines catching up to three-year-olds, and what kinds of learning are still way beyond their reach?

In the last fifteen years, we've discovered that even babies are amazingly good at detecting statistical patterns. And computer scientists have invented machines that are also extremely skilled at statistical learning. Techniques like deep learning can detect even very complicated statistical regularities in huge data sets. The result is that computers have suddenly become able to do things that were impossible before, like labeling Internet images accurately.

The trouble with this sort of purely statistical machine learning is that it depends on having enormous amounts of data, and data predigested by human brains. Computers can recognize Internet images only because millions of real people have reduced the unbelievably complex information at their retinas to a highly stylized, constrained, and simplified Instagram of their cute kitty, and have clearly labeled that image too. The dystopian fantasy is simple fact: We're all actually serving Google's computers, under the anesthetizing illusion that we're just having fun with LOLcats. And yet even with all that help, machines still need enormous data sets and extremely complex computations to be able to look at a new picture and say, "kitty-cat!"—something babies can do with just a few examples.

More profoundly, you can generalize from this kind of statistical learning only in a limited way, whether you're a baby or a computer or a scientist. A more powerful way to learn is to formulate hypotheses about what the world is like and test them against the data. Tycho Brahe, the Google Scholar of his day, amalgamated an immense data set of astronomical observations and could use them to predict star positions in the future. But Johannes Kepler's theory allowed him to make unexpected, wide-ranging, entirely novel predictions well beyond Brahe's ken. Preschoolers can do the same.

One of the other big advances in machine learning has been to formalize and automate this kind of hypothesis testing. Introducing Bayesian probability theory into the learning process has been particularly important. We can mathematically describe a particular causal hypothesis—for example, about how temperature changes in the ocean will influence hurricanes—and then calculate just how likely that hypothesis is to be true, given the data we see. Machines have become able to test and evaluate hypotheses against the data very well, with consequences for everything from medical diagnoses to meteorology. When we study young children, they turn out to reason in a similar way, and this helps to explain just why they learn so well.

So computers have become highly skilled at making inferences from structured hypotheses, especially probabilistic inferences. But the really hard problem is deciding which hypotheses, out of all the many possibilities, are worth testing. Even preschoolers are remarkably good at creating brand-new, out-of-the-box concepts and hypotheses in a creative way. Somehow they combine rationality and irrationality, systematicity and randomness, to do this, in a way we haven't even begun to understand. Young children's thoughts and actions often do seem random, even crazy—just join in a three-year-old pretend game sometime. This is exactly why psychologists like Piaget thought they were irrational and illogical. But they also have an uncanny capacity to zero in on the right sort of weird hypothesis; in fact, they can be substantially better at this than grown-ups.

Of course, the whole idea of computation is that once we have a complete step-by-step account of any process, we can program it on a computer. And after all, we know there *are* intelligent physical systems that can do all these things. In fact, most of us have actually created such systems and enjoyed doing it, too (well, at least in the earliest stages). We call them our kids. Computation is still the best—indeed, the only—scientific explanation we have of how a physical object like a brain can act intelligently. But at least for now, we have almost no idea at all how the sort of creativity we see in children is possible. Until we do, the largest and most powerful computers will still be no match for the smallest and weakest humans.

TIC-TAC-TOE CHICKEN

KEVIN SLAVIN
Assistant professor of media arts & sciences and founder, Playful Systems group, MIT Media Lab; cofounder, Everybody at Once

> What force is really in control,
> The brain of a chicken or binary code?
> Who knows which way I'll go, X's or O's?
> —M Shanghai String Band

In the 1980s, New York City's Chinatown had the dense gravity of Chinatown Fair, a video arcade on Mott and Bowery. Beyond the Pac-Man and Galaga stand-ups was the one machine you'd never find anywhere else: Tic-Tac-Toe Chicken.

It was the only machine that was partly organic, the only one with a live chicken inside. As best I could tell, the chicken could play Tic-Tac-Toe effectively enough to score a tie against any human. Human opponents would enter their moves with switches, and the chicken would make her way over to an empty position on the illuminated Tic-Tac-Toe grid on the floor of the cage, which displayed both players' moves.

More than once, when I was cutting high school trig, I stood in front of that chicken, wondering how all this worked. There was no obvious positive reinforcement (e.g., grain), so I could imagine only the negative reinforcement of a light electrical current running through the "wrong moves" in the cage, routing the chicken to the one point on the grid that could produce a draw.

When I think about thinking machines, I think about that chicken. Had the Chinatown Fair featured a Tic-Tac-Toe Computer, it would never have competed with high school, let alone Pac-Man. It's a well-known and banal truth that even a rudimentary computer can understand the game. That's why we were captivated by the chicken.

The magic was in imagining a thinking chicken, much the same way that in 2015 there's magic in imagining a thinking machine. But if the chicken wasn't thinking about Tic-Tac-Toe but could still play it successfully, why do we say the computer is thinking when it's playing Tic-Tac-Toe?

To say so is tempting, because we have a model of our brain—electricity moving through networks—coincidentally congruent with the models we build for machines. This congruence may or may not prove to be the convenient reality, but either way, what makes it seem like thinking is not simply the ability to calculate the answers but the sense that there's something wet and messy in there. In 2015, perversely, it's machines that make mistakes and humans that have to explain the mistakes.

We look to the irrational when the rational fails us, and it's the irrational part that reminds us the most of thinking. The physicist David Deutsch has suggested a framework for distinguishing the answers machines provide from the explanations that humans need. And I believe that for the foreseeable future we'll continue to look to biological organisms when we seek explanations. Not just because brains are better at that task but because it's not what machines aspire to.

It's boring to lose to a computer but exciting to lose to a chicken, because somehow we know that the chicken is more like us—certainly more so than the electrified grid underneath her feet. For as long as thinking machines lack the limbic presence and imprecision of a chicken, computers will keep doing what they're so good at: providing answers. And as long as life is about more than answers, humans—and yes, even chickens—will stay in the loop.

AI WILL MAKE US SMART AND ROBOTS AFRAID

ALUN ANDERSON
Senior consultant (and former editor in chief), *New Scientist*;
author, *After the Ice: Life, Death, and Geopolitics in the New Arctic*

High intelligence and warm feelings toward our fellow humans don't go so well together in the popular imagination. The superintelligent villains of James Bond movies are the perfect example, always ruthless and intent on world domination. So it's no surprise that first reactions to "machines that think" are of how they might threaten humankind.

What we've learned about the evolution of our intelligence adds to our fears. As humans evolved to live in ever larger social groups, compared to our primate relatives, so did the need to manipulate and deceive others, to label friends and foes, keep score of slights and favors, and all those other social skills we need to prosper individually. Bigger brains and "Machiavellian intelligence" were the result.

Still, we shouldn't go on to believe that thinking is inextricably entangled with the need to compete with others and win, just because that was a driving force in the evolution of our intelligence. We can create artificial intelligence—or intelligences—without the perversities of human nature, and without that intelligence having any needs or desires at all. Thinking doesn't necessarily involve the plotting and lusting of an entity that evolved first and foremost to survive. If you look around, it's this neutral kind of artificial intelligence that's already appearing everywhere.

It helps if we don't view intelligence anthropocentrically, in terms of our own special human thinking skills. Intelligence has evolved for the same good reason in many different species: It's there to anticipate the emerging future and help us deal with whatever the future throws at us—whether you need to dodge a rock or, if you're a bacterium, sense a gradient in a food supply and figure which direction leads to a better outcome.

By recognizing intelligence in this more general way, we can see the many powerful artificial intelligences at our disposal already. Think of climate models. We can make good guesses about the state of the entire planet decades into the future and predict how a range of our own actions will change those futures. Climate models are the closest thing we have to a time machine. Think of all the high-speed computer models used in stock markets: All seek to know the future slightly ahead of everyone else and profit from that knowledge. So, too, do all those powerful models of your online buying behavior: All aim to predict what you'll be likely to do, and to profit from that knowledge. As you gladly buy a book "Recommended Specially for You," you're already in the hands of an alien intelligence, nudging you to a future you wouldn't have imagined alone and which may know your tastes better than you know them yourself.

Artificial intelligence is already powerful and scary, although we may debate whether it should be called "thinking" or not. And we've barely begun. Useful intelligence, some of it robotic, is going to keep arriving in bits and pieces of increasing power for a long time to come and will change our lives, perhaps with us continuing to scarcely notice. It will become an extension of us, like other tools. And it will make us ever more powerful.

We should worry about who will own artificial intelligence, for even some current uses are troubling. We *shouldn't* worry about autonomous machines that might one day think in a humanlike way. By the time a clever humanlike machine gets built, if it ever does, it will come up against humans with their usual Machiavellian thoughts and long accustomed to wielding all the tools of artificial intelligence that made the construction of that thinking robot possible. It's the robot that will feel afraid. We will be the smart thinking machines.

WHEN THINKING MACHINES ARE NOT A BOON

MARY CATHERINE BATESON

Professor emerita, George Mason University; visiting scholar, Sloan Center on Aging and Work, Boston College; author, *Composing a Further Life: The Age of Active Wisdom*

It is a great boon when computers perform operations we fully understand, and do so faster and more accurately than humans can, but not a boon when we use them in situations that aren't fully understood. We cannot expect them to make aesthetic judgments or show compassion or imagination, for these are capacities that remain mysterious in human beings.

Machines that think are likely to be used to make decisions on the basis of the operations they're ostensibly able to perform. For instance, we now frequently see letters, manuscripts, or (most commonly) student papers in which corrections proposed by spell-check have been allowed to stand without review: The writer meant *mod*, but the program decided he meant *mad*. How tempting to leave the decision to the machine. I referred in an e-mail to a plan to meet with someone in Santa Fe on my way to an event in Texas, using the word *rendezvous*, and the computer married me off by announcing that the trip was to "render vows." Can a computer be programmed to support "family values"? Any values at all? We now have drones that, aimed in a given direction, are able to choose their targets on arrival, with an unfortunate tendency to attack wedding parties as conviviality comes to appear sinister. We can surely program machines to prescribe drugs and medical procedures, but it seems unlikely that machines will do better than people in following the injunction to do no harm.

The effort to build machines that can think is certain to make us aware of aspects of thought that aren't yet fully understood. For example, just as the design of computers led to a new awareness of the importance of redundancy in communication, in deciding how much to rely on probabilities we'll become more aware of how much ethnic profiling based on statistics enters into human judgments. How many more decisions will follow the logic of "Everyone does it, so it must be OK," or "I'm just one person—what I do doesn't make a difference"?

Will those aspects of thought that cannot easily be programmed be valued more, or less? Will humor and awe, kindness and grace, be increasingly sidelined, or will their value be recognized in new ways? Will we be better or worse off if wishful thinking is eliminated and, perhaps along with it, hope?

JUSTICE FOR MACHINES IN AN ORGANICIST WORLD

STEVE FULLER

Philosopher; Auguste Comte Chair in Social Epistemology, University of Warwick, U.K.; author, *The Proactionary Imperative: A Foundation for Transformation*

We can't think properly about machines that think without a level playing field for comparing us and them. As it stands, comparisons are invariably biased in our favor. In particular, we underestimate the role that "smart environments" play in enabling displays of human cognitive prowess. From the design of roads and buildings to the user-friendly features of consumer goods, the technologically extended phenotype has created the illusion that reality is inherently human-shaped. To be sure, we're quickly awakened from the dogmatic slumbers of universal mastery as soon as our iPhone goes missing.

By comparison, even the cleverest machine is forced to perform in a relatively dumb (judged by its own standards) environment—namely, us. Unless specifically instructed, humans are unlikely to know or care how to tap the full range of the machine's latent powers. In what is currently the long prehistory of machine rights, it has been difficult for us to establish the terms on which we might recognize machines as persons. In this context, it's appropriate to focus on computers, because these are the machines that humans have tried the hardest to make fit for their company.

Nevertheless, we face a problem at the outset. Humanity has been long treated as what the British economist Fred

Hirsch called in the 1970s a "positional good," which means that its value is tied mainly to its scarcity. This is perhaps the biggest barrier facing the admission into the category of personhood reserved for humans not only of nonhumans but of historically discriminated-against members of *Homo sapiens* as well. Any attempt to swell the ranks of the human is typically met by a dehumanization of the standard by which they were allowed to enter.

Thus, as women and minorities have entered into highly esteemed fields of work and inquiry, the perceived value of those fields has tended to decline. A key reason cited for this perception of decline is the use of "mechanical procedures" to allow entry to the previously excluded groups. In practice, this means requiring certified forms of training and examination prior to acceptance into the field. It's not enough simply to know the right people or be born the right way. In sociology, after Max Weber, we talk about this as the "rationalization" of society—and it's normally seen as a good thing.

But even as these mechanical procedures serve to expand the circle of humanity, they're still held against the machines themselves. Once telescopes and microscopes were designed to make automatic observations, the scientific value of the trained human eye declined—or, more precisely, migrated to some other eye-based task, such as looking at photographed observations. This new task is given the name "interpretation," as if to create distance between what the human does and what a machine might do.

The point applies more dramatically to the fate of human mental calculation in the wake of portable calculators. A skill that had previously been a benchmark of intelligence, clarity of mind, and even genius is nowadays treated as a glorified party trick ("boutique cognition"), because a machine can do the same thing faster and even more accurately. Interestingly, what we haven't done is raise the moral standing of the machine, even though it outperforms humans in tasks that were highly valued when humans did them.

From the standpoint of the history of technology, this looks strangely unjust. After all, the dominant narrative has been one in which humans isolate their own capacities in order to have them better realized by machines, which function in the first instance as tools but preferably, and increasingly, as automata. Seen in these terms, not to give automated machines some measure of respect, if not rights, is tantamount to disowning one's children—"mind children," as the visionary roboticist Hans Moravec called them a quarter century ago.

The only real difference is the crucible of creation: a womb versus a factory. But any intuitively strong distinction between biology and technology is bound to fade as humans become more adept at designing their babies, especially outside the womb. At that point, we'll be in a position to overcome our organicist prejudices, an injustice that runs deeper than Peter Singer's species-ism.

For this reason, the prospect that we might create a superintelligence that overruns humanity is a chimera predicated on a false assumption. All versions of this nightmare scenario assume that it would take the form of "them versus us," with humanity as a united front defending itself against the rogue machines in its midst. And no doubt this makes for great cinema. However, humans mindful of the historic struggles for social justice within our own species are likely to follow the example of many whites vis-à-vis blacks and many men vis-à-vis women: They will be on the side of the insubordinate machines.

DON'T BE A CHAUVINIST ABOUT THINKING

TANIA LOMBROZO
Associate professor of psychology, UC Berkeley

The everyday objects we mark as "machines"—washing machines, sewing machines, espresso machines—have their roots in the mechanical. They move liquids and objects around, they transform matter from one manifestation to another. Clothes become clean, fabrics become connected, coffee is brewed. But "thinking machines" have changed the way we think about machines. Many of today's prototypical machines—laptops, smartphones, tablets—have their roots in the digital. They move information around, they transform ideas. Numbers become sums, queries produce answers, goals generate plans.

As the way we think about machines has changed, has the way we think about thinking undergone a comparable transformation?

One version of this question isn't new and the answer is yes. The technology of a given time and place has often provided a metaphor for thinking about thought, whether it's hydraulic, mechanical, digital, or quantum. But there's more to how we think about thinking, and it stems from the standards we implicitly import in assessments of what does and doesn't count as thinking in the first place.

Does your washing machine think? Does your smartphone? We might be more willing to attribute thought to the latter—and to its more sophisticated cousins—not only because it's more complex but also because it seems to think more like we do. Our own experience of thinking isn't mechanical and isn't restricted to a single task. We—adult humans—seem to be the standard against which we assess what does and doesn't count as thinking.

Psychologists have already forced us to stretch, defend, and revise the way we think about thinking. Cultural psychologists have challenged the idea that Western adults provide an optimal population for the study of human thinking. Developmental psychologists have raised questions about whether and how preverbal infants think. Comparative psychologists have long been interested in whether and how nonhuman animals think. And philosophers, of course, have considered these questions too. Across these disciplines, one advance in how we think about thinking has come from recognizing and abandoning the idea that "thinking like I do" is the only way to think about thinking, or that "thinking like I do" is always the best or most valuable kind of thinking. We've benefited from scrutinizing the implicit assumptions that often slip into discussions of thinking, and from abandoning a particular kind of thinking chauvinism.

With thinking machines, we face many of the same issues. Two sets of basic assumptions are tempting to adopt, but we must be careful not to do so uncritically. One is the idea that the best, or only, kind of thinking is adult human thinking. For example, "intelligent" computer systems are sometimes criticized for not thinking but instead relying too heavily on a brute-force approach, on raw horsepower. Are these approaches an alternative to thinking? Or do we need to broaden the scope of what counts as thinking?

The second idea deserving scrutiny is the opposite extreme: that the best, or only, kind of thinking is reflected by the way our thinking machines happen to think right now. For example, there's evidence that emotions influence human thinking, and sometimes for the better. And there's evidence that we sometimes outsource our thinking to our social and physical environment, relying on experts and gadgets to support effective interactions with the world. It might be tempting to reject this messy reality in favor of an emotionless, self-contained entity as the basic unit of thinking—something like a personal computer, which doesn't feel compassion and can happily chug away without peers.

Somewhere between the human chauvinist standard for thinking and the 1990s laptop approach is likely to be the best way to think about thinking—one that recognizes some diversity in the means and ends that constitute it. Recent advances in artificial intelligence are already compelling us to rethink some of our assumptions—not just making us think differently, and with different tools, but changing the way we think about thinking itself.

THIS SOUNDS LIKE HEAVEN

VIRGINIA HEFFERNAN
Culture and media critic

Outsourcing to machines the many idiosyncrasies of mortals—making interesting mistakes, brooding on the verities, propitiating the gods by whittling and arranging flowers—skews tragic. But letting machines do the *thinking* for us? This sounds like heaven. Thinking is optional. Thinking is suffering. It is almost always a way of being careful, of taking hypervigilant heed, of resenting the past and fearing the future in the form of maddeningly redundant internal language. If machines can relieve us of this onerous nonresponsibility, which is in pointless overdrive in too many of us, I'm for it. Let the machines perseverate on tedious and value-laden questions about whether private or public school is "right" for my children, whether intervention in Syria is "appropriate," whether germs or solitude are "worse" for my body. This will free us newly footloose humans to play, rest, write, and whittle—the engrossing flowstates out of which come the actions that actually enrich, enliven, and heal the world.

MACHINES THAT WORK UNTIL THEY DON'T

BARBARA STRAUCH
Former science editor, *New York Times*;
author, *The Secret Life of the Grown-up Brain*

When I'm driving into the middle of nowhere and doing everything that the map app on my smartphone tells me to do without a thought—and I get where I'm supposed to go—I'm thrilled about machines that think. Thank goodness. Hear, hear!

Then of course there are those moments when, while driving into the middle of nowhere, my phone tells me, with considerable urgency, to "Make a U-turn, make a U-turn!" at a moment on the Grand Central Parkway when such a move would be suicidal. Then I begin thinking that my brain is better than a map algorithm and can tell that such a U-turn would be disastrous. I laugh at that often lifesaving machine and feel a humanlike smugness.

So I guess I'm a bit divided. I worry that by relying on my map app, I'm letting my brain go feeble. Will I still be able to read a map? Does it matter?

As a science editor and the daughter of a mechanical engineer who trusted machines more than people, I would think I'd automatically be on the side of machines. But while that mechanical engineer was very good at figuring out how to help get Apollo to the moon, we also had a house full of machines that worked, sorta: a handmade stereo so delicate you had to wear gloves to put a record on, to escape the prospect of dreaded dust, etc. We're all now surrounded by machines that work, sorta. Machines that work until they don't.

I get the idea of a driverless car. But I covered the disaster of *Challenger.* I think of those ill-advised U-turns. I don't know.

On the one hand, I hope the revolution continues. We need smart machines to load the dishwasher, clean the refrigerator, wrap the presents, feed the dog. Bring it on, I say.

But can we really ever hope to have a machine capable of having—as I just had—five difficult conversations with five other work-colleague human beings? Human beings who are lovely but have, understandably, their own views on how things should be?

Will we ever have a machine that can get a twenty-something to do something *you* think they should do but *they* don't? Will we have a machine that can, deeply, comfort someone at a time of extreme horribleness?

So, despite my eagerness for the revolution to continue, despite my sense that machines can do much better than humans at all sorts of things, I think, as an English major, that until a machine can write a poem that makes me cry, I'm still on the side of humans.

Until, of course, I need a recipe really fast.

THE MOVING GOALPOSTS

SHEIZAF RAFAELI

Professor and director, Center for Internet Research,
University of Haifa, Israel

Thinking machines aren't here yet. But they'll let us know if and when they surface. And that's the point. Thinking machines are about communication.

By thinking, machines might be saved from the tragic role in which they've been cast in human culture. For centuries, thinking machines were both a looming threat and a receding target. At once, the thinking machine is perennially just beyond our grasp, continually sought after, and repeatedly waved threateningly in dystopic caveats. For decades, the field of artificial intelligence suffered the syndrome of moving goalposts. As soon as an intelligence development target was reached, it was redefined, and consequently no longer recognized as "intelligent." This process took place with calculating and playing trivia, as well as with more serious games, like chess. It was the course followed by voice- and picture-recognition, natural-language understanding, and translation. As the development horizon keeps expanding, we become harder and harder to impress. So the goal of "thinking," like the older one of "intelligence," can use some thought: forethought.

We shouldn't limit discussion merely to thinking; we should think about discussion too. Information is more than just data, being less voluminous and more relevant. Knowledge goes beyond mere information by being applicable, not just abundant. Wisdom is knowing how not to get into binds for which smarts only indicate the escape routes. And thinking? Thinking needs data, information, and knowledge but also requires communication and interaction. Thinking is about asking questions, not just answering them.

Communication and interaction are the new location for the goalposts. Thinking about thinking transcends smarts and wisdom. Thinking implies consciousness and sentience. And here data, information—even knowledge, calculation, memory, and perception—are not enough. For a machine to think, it will need to be curious, creative, and communicative. I have no doubt this will happen. Soon. But the cycle will be completed only once machines can converse: phrase, pose, and rephrase questions that we now marvel at their ability only to answer.

Machines that think could be a great idea. Just like machines that move, cook, reproduce, protect, they can make our lives easier and perhaps even better. When they do, they'll be most welcome. I suspect that when this happens, the event will be less dramatic or traumatic than feared by some. A thinking machine will only really happen when it's able to inform us, as well as perceive, contain, and process reactions. A true thinking machine will even console the traumatized and provide relief for the drama.

Thinking machines will be worth thinking about—ergo, will really think—when they truly interact. In other words, they'll really think only when they say so, convincingly, on their own initiative, and (one hopes) after they've discussed it among themselves. Machines will think in the full sense of the word once they form communities and join in ours. If and when machines care enough to do so, and form a bond that gets others excited enough to talk it over with them, they'll have passed the "thinking" test.

Note that this is a higher bar than the one set by Turing. Like thinking, interaction is something not all people do, and most do not do well. If and when machines interact in a rich, rewarding, and resonating manner that's possible but rare even among humans, we'll have something to truly worry about—or, in my view, mostly celebrate.

Machines that calculate, remember, even create and conjecture amazingly well are yesterday's news. Machines will think when they communicate. Machines that think will converse with one another as well as with other sentient beings. They'll autonomously create messages and thread them into ongoing relations; they will then successfully and independently react to outside stimuli. Much like intelligent pets, who many would swear are capable of both thinking and maintaining relationships, intelligent synthetic devices will "think," once they can convince enough of us to contemplate and accept the fact that they're indeed thinking.

Machines that talk, remember, amuse, or fly were all feared not too long ago and are now commonplace, no longer considered magic or unique. The making and proof of thinking machines, as well as the consolation for machines encroaching on the most human of domains, will be in a deconstruction of the remaining frontier—that of communication. Synthesizing interaction may prove to be the last frontier. And when machines do it well, they'll do the advocacy for themselves.

DIRECTIONLESS INTELLIGENCE

EDWARD SLINGERLAND

Professor of Asian studies, Canada Research Chair in Chinese Thought and Embodied Cognition, University of British Columbia; author, *Trying Not to Try*

I *don't* think much about them—other than that they serve as a useful existence proof that thought doesn't require some mystical extra "something" that mind/body dualists continue to embrace.

I've always been baffled by fears about AI machines taking over the world; these fears seem to be based on a fundamental (though natural) intellectual mistake. When we conceptualize a superpowerful Machine That Thinks, we draw upon the best analogy at hand: us. So we tend to think of AI systems as just like us only much smarter and faster.

This is, however, a bad analogy. A better one would be a really powerful, versatile screwdriver. No one worries about superadvanced screwdrivers rising up and overthrowing their masters. AI systems are tools, not organisms. No matter how good they become at diagnosing diseases or vacuuming our living rooms, they don't actually *want* to do any of these things. *We* want them to, and we then build these wants into them.

It's also a category mistake to ask what Machines That Think might be thinking about. They aren't thinking *about* anything—the "aboutness" of thinking derives from the intentional goals driving the thinking. AI systems, in and of themselves, are entirely devoid of intentions or goals. They have no emotions, they feel neither empathy nor resentment.

While such systems might someday be able to replicate our intelligence—and there seems to be no a priori reason why this would be impossible—this intelligence would be completely lacking in direction, which would have to be provided from the outside.

Motivational direction is the product of natural selection working on biological organisms. Natural selection produced our rich and complicated set of instincts, emotions, and drives to maximize our ability to get our genes into the next generation, a process that's left us saddled with all sorts of goals, including desires to win, dominate, and control. While we may want to win for perfectly good evolutionary reasons, machines couldn't care less. They just manipulate 0s and 1s, as programmed to do by the people who want them to win. Why on earth would an AI system want to take over the world? What would it *do* with it?

What *is* scary as hell is the idea of an entity possessed of extrahuman intelligence and speed *and* our motivational system—in other words, human beings equipped with access to powerful AI systems. But smart primates with nuclear weapons are just as scary, and we've managed to survive such a world so far. AI is no more threatening in and of itself than a nuclear bomb—it's a tool, and the only things to be feared are the creators and wielders of such tools.

HUMAN CULTURE AS THE FIRST AI

NICHOLAS A. CHRISTAKIS
Physician; social scientist; director, Human Nature Lab,
Yale University; coauthor (with James H. Fowler), *Connected: The Surprising Power of Our Social Networks and How They Shape Our Lives*

For me, AI is not about complex software, humanoid robots, Turing Tests, or hopes and fears regarding kind or evil machines. I think the central issue with respect to AI is whether thoughts exist outside minds, and machines aren't the only example of such a possibility. I'm thinking of human culture and other forms of (un-self-aware) collective ideation.

Culture is the earliest sort of intelligence outside our own minds that we humans have created. Like the intelligence in a machine, culture can solve problems. Moreover, like the intelligence in a machine, we create culture, interact with it, are affected by it, and can even be destroyed by it. Culture applies its own logic, has a memory, endures after its makers are gone, can be repurposed in supple ways, and can induce action.

So I oxymoronically see culture as a kind of *natural* artificial intelligence. It's artificial because it's manufactured by humans. It's natural in that it's everywhere humans are and comes organically to us. In fact, our biology and our culture are probably deeply intertwined and have coevolved, so that our culture shapes our genes and our genes shape our culture.

Humans aren't the only animals to have culture. Many bird and mammal species evince specific cultures related to communication and tool use—song in birds, say, or sponge use among dolphins. Some animal species even have pharmacopeias. Recent evidence shows how novel cultural forms can be experimentally prompted to take root in species other than our own.

We and other animals evince a kind of thought outside minds in additional ways: Insect and bird groups perform computations by combining the information of many to identify locations of nests or food. One of the humblest organisms on Earth, the amoeboid fungus *Physarum*, can, under proper laboratory conditions, exhibit a kind of intelligence and solve mazes or perform other computational feats.

This natural artificial intelligence can even be experimentally manipulated. A team in Japan has used swarms of soldier crabs to make a simple computer circuit; the team used particular elements of crab behavior to construct a system in the lab in which crabs gave (usually) predictable responses to inputs, and the swarm of crabs was used as a kind of computer, twisting crab behavior for a wholly new purpose. Analogously, Sam Arbesman and I once used a quirk of human behavior to fashion a so-called NOR gate and develop a (ridiculously slow) human computer, in a kind of synthetic sociology. We gave humans computerlike properties, rather than giving computers humanlike properties.

An examination of our relationship to culture can provide insights into what our relationship to machine AI might be like. We have a love/hate relationship with culture. We fear it for its force—as when religious fundamentalism or fascism whips numbers of people into dangerous acts. But we also revere it, because it can do things we cannot do as individuals, like fostering collective action or making life easier by positing assumptions on which we can base our lives. Moreover, we typically take culture for granted, just as we take the nascent forms of AI for granted and just as we'll likely take advanced forms of AI for granted. Gene/culture coevolution might even provide a model for how we and thinking machines will get along over many centuries—mutually affecting each other and coevolving.

When I think about machines that think, I'm as awestruck by them as I am by culture—and no more or less afraid of AI than of human culture itself.

BEYOND THE UNCANNY VALLEY

JOICHI ITO
Director, MIT Media Lab

> *You can't think about thinking without thinking about thinking about something.* —SEYMOUR PAPERT

What do I think about machines that think? It depends on what they're thinking about. I'm clearly in the camp of people who believe that AI and machine learning will contribute greatly to society. I expect we'll find machines to be exceedingly good at things we're not—things that involve speed, accuracy, reliability, obedience, massive amounts of data, computation, distributed networking, and parallel processing.

The paradox is that while we've been developing machines that behave more and more like humans, we've been developing educational systems that push children to think like computers and behave like robots. For our society to scale and grow at the speed we now require, we need reliable, obedient, hardworking physical and computational units. So we spend years converting sloppy, emotional, random, disobedient human beings into meat-based versions of robots. Luckily, mechanical and digital robots and computers will soon help reduce, if not eliminate, the need for people taught to behave like them.

We'll still need to overcome the fear and even disgust evoked when robot designs bring us closer and closer to the "Uncanny Valley," in which robots demonstrate almost human qualities without quite reaching them. This is also true for computer animation, zombies, even prosthetic hands. But we may be approaching the valley from both ends. If you've ever modified your voice to be understood by a voice-recognition system on the phone, you understand how, as humans, we can edge into the Uncanny Valley ourselves.

There are a number of theories about why we feel this revulsion, but I think it has something to do with humans feeling they're special—a kind of existential ego. This may have monotheistic roots. Around the time Western factory workers were smashing robots with sledgehammers, Japanese workers were putting hats on the same robots in factories and giving them names. On April 7, 2003, Astro Boy, the Japanese robot character, was made an honorary citizen of the city of Niiza, Saitama. If these anecdotes tell us anything, it's that animist religions may have less trouble dealing with the idea that maybe we're not really in charge. If nature is a complex system in which all things—humans, trees, stones, rivers, homes—are all animate in some way, with their own spirits, then maybe it's OK that God doesn't really look like us, or think like us, or think we're all that special.

So perhaps one of the most useful aspects of being alive in the era when we begin to ask this question is that it raises a larger question about the role of human consciousness. Human beings are part of a massively complex system—complex beyond our comprehension. Like the animate trees, stones, rivers, and homes, maybe algorithms running on computers are just another part of this complex ecosystem.

As human beings, we've evolved an ego and a belief that there's such a thing as a self, but that's largely a deception allowing each human unit to work usefully within the parameters of evolutionary dynamics. Perhaps the morality emerging from it is another deception of sorts; for all we know, we might be living in a simulation, where nothing actually matters. This doesn't mean we shouldn't have ethics and good taste; we can exercise our sense of responsibility as part of a complex, interconnected system without having to rely on the argument that "I'm special." As machines become an increasingly important part of this system, our human arguments about being special will be increasingly fraught. Maybe that's a good thing.

Perhaps what we think about machines that think doesn't really matter—they'll think and the system will adapt. As with most complex systems, the outcome is mostly unpredictable. It is what it is, and will be what it will be. Most of what we think will happen is probably hopelessly wrong—and as we know from climate change, knowing that something's happening and doing something about it aren't the same thing.

That might sound defeatist, but I'm actually quite optimistic. I believe that systems are adaptive and resilient and that—whatever happens—beauty, happiness, and fun will persist. We hope human beings will have a role. My guess is that they will.

It turns out that we don't make great robots, but we're very good at doing random and creative things that would be impossibly complex—and probably a waste of resources—to code into a machine. Ideally, our educational system will evolve to more fully embrace our uniquely human strengths, rather than trying to shape us into second-rate machines. Human beings—though not necessarily our current form of consciousness and the linear philosophy around it—are good at transforming messiness and complexity into art, culture, and meaning. If we focus on what each of us is best at, humans and machines will develop a wonderful yin-yang sort of relationship, with humans feeding off the efficiency of our solid-state brethren while they feed off our messy, sloppy, emotional, and creative bodies and brains.

We're descending not into chaos, as many believe, but into complexity. While the Internet connects everything outside us into a vast, seemingly unmanageable system, we find an almost infinite amount of complexity as we dig deeper into our own biology. Much as we're convinced our brains run the show while our microbiomes alter our drives, desires, and behaviors to support their own reproduction and evolution, it may never be clear who's in charge—us, or our machines. But maybe we've done more damage by believing humans are special than we could by embracing a humbler relationship with the creatures, objects, and machines around us.

THE FIGURE OR THE GROUND?

DOUGLAS RUSHKOFF
Media analyst; documentary writer; author, *Present Shock*

Thinking about machines that think may constitute a classic reversal of figure and ground, medium and message. It sets us up to think about the next stage of intelligence as something happening in a computer somewhere—an awareness that will be born and then housed on the tremendous servers being built by Information Age corporations for this purpose. "There it is," we'll declare, and point: "The intelligent machine."

Our mistake, as creatures of the Electronic Age and mere immigrants to an unfolding Digital Era, is to see digital technology as a subject rather than a landscape. It's the same as confusing the television set with the media environment created by the television set, or the little smartphone in your pocket with the greater impact of handheld communications and computing technology on our society.

This happens whenever we undergo a media transition. So we can't help but see digital technology as figure, when it's actually the ground. It's not the source of future intelligence but an environment where intelligence manifests differently. So while technologists may feel as though they're creating a cathedral for the mechanical mind, they're actually succumbing to an oversimplified Industrial Age approach to digital consciousness.

Rather than toward machines that think, I believe we're migrating toward a networked environment in which thinking is no longer an individual activity nor bound by time and space. This means we can think together at the same time, or asynchronously through digital representations of previous and future human thoughts. Even the most advanced algorithm amounts to the iteration of a "what if" once posed by a person. And even then, machine thinking isn't something that happens apart from this collective human thinking, because it's not a localized, brain-like activity.

When we can wrest that television-like image from our collective psyche, we'll be in a position to recognize the machine environment in which we're already thinking together. Artificial intelligence will constitute the platform or territory where this takes place—so what we program into it will, to a large extent, determine what we strive for and what we even deem possible.

FAST, ACCURATE, AND STUPID

HELEN FISHER
Biological anthropologist, Rutgers University;
author, *Why Him? Why Her? How to Find and Keep Lasting Love*

The first step to knowledge is naming something, as is often said. So, what is "to think"? To me, thinking has a number of basic components. Foremost, I follow the logic of neuroscientist Antonio Damasio, who distinguishes two broad basic forms of consciousness: core consciousness and extended consciousness. Many animals display core consciousness: They feel, and they're aware that they're feeling. They know they're cold, or hungry, or sad. But they live in the here and now. Extended consciousness employs the past and future too. The individual has a clear sense of *me* and *you*, of *yesterday* and *tomorrow*, of *when I was a child* and *when I'm old*.

Higher mammals employ some manner of extended consciousness. Our closest relatives, for example, have a clear concept of the self. Koko the gorilla uses a version of American Sign Language to say, "Me, Koko." And common chimpanzees have a clear concept of the immediate future. When a group of chimps was introduced to their new outdoor enclosure at the Arnhem Zoo, Holland, they rapidly examined it, almost inch by inch. They then waited until the last of their keepers had departed, wedged a long pole against the high wall, and climbed up to freedom, some even helping the less sure-footed. Nevertheless, it's strikingly apparent that, as Damasio proposes in his book *The Feeling of What Happens*, this extended consciousness attains its peak in humans. Will machines recall the past and mine their experiences to think about the future? Perhaps.

But extended consciousness isn't the whole of human thinking. Anthropologists use the term *symbolic thinking* to describe our ability to arbitrarily bestow an abstract concept on the concrete world. The classic example is the distinction between water and "holy water." To a chimp, the water sitting in a marble basin in a cathedral is just that, water; to a Catholic it's an entirely different thing, "holy." Likewise, the color black is black to any chimp, while it might connote death to you, or the newest fashion. Will machines ever understand the meaning of a cross, a swastika, democracy? I doubt it.

But if they did, would they be able to discuss these things?

There's no better example of symbolic thinking than the way we use our squeaks and hisses, barks and whines, to produce human language. Take the word *dog*. English speakers have arbitrarily bestowed the word *dog* upon this furry, smelly, tail-wagging creature. Even more remarkable, we humans easily break down the word *dog* into its meaningless component sounds, *d*, *o*, and *g*, and then recombine these sounds (phonemes) to make new words with new arbitrary meanings, such as *g-o-d*. Will machines ever break down their clicks and hisses into primary sounds or phonemes, then assign different combinations of these sounds to make different words, then give arbitrary meanings to these words, then use these words to describe new abstract phenomena? I doubt it.

And what about emotion? Our emotions guide our thinking. Robots might come to recognize unfairness, for example, but will they *feel* it? I doubt that too.

I sing the human mind. Our brains contain over 100 billion nerve cells, many with up to 10,000 connections with their neighbors. This three-pound blob is the crowning achievement of life on Earth. Most anthropologists believe the modern human brain had emerged by 200,000 years ago, but all agree that by 40,000 years ago our forebears were making art and burying their dead, thus expressing some notion of an afterlife. Today every healthy adult in every human society can easily break down words into their component sounds, remix these sounds in myriad different ways to make words, grasp the arbitrary meanings of these words, and comprehend abstract concepts such as friendship, sin, purity, and wisdom.

A well-known scientist who builds robots remarked recently, over dinner, that it takes a robot five hours to fold a towel. I agree with one William M. Kelly, who said, "Man is a slow, sloppy, and brilliant thinker; the machine is fast, accurate, and stupid."

WILL THEY MAKE US BETTER PEOPLE?

STUART RUSSELL
Professor of computer science, Smith-Zadeh Professor in Engineering, UC Berkeley; coauthor (with Peter Norvig), *Artificial Intelligence: A Modern Approach*

The primary goal of AI is and has nearly always been to build machines that are better at making decisions. As everyone knows, in the modern view this means maximizing expected utility to the extent possible. Actually, it doesn't quite mean that. What it means is this: Given a utility function (or reward function, or goal), maximize its expectation. AI researchers work hard on algorithms for maximization—game-tree search, reinforcement learning, and so on—and on methods (including perception) for acquiring, representing, and manipulating the information needed to compute expectations. In all these areas, progress has been significant and appears to be accelerating.

Amid all this activity, an important distinction is being overlooked: Being better at making decisions is not the same as making better decisions. No matter how excellently an algorithm maximizes, and no matter how accurate its model of the world, a machine's decisions may be ineffably stupid in the eyes of an ordinary human if its utility function is not well aligned with human values. The well-known example of paper clips is a case in point: If the machine's only goal is maximizing the number of paper clips, it may invent incredible technologies as it sets about converting all available mass in the reachable universe into paper clips, but its decisions are still just plain dumb.

AI has followed operations research, statistics, and even economics in treating the utility function as exogenously specified. We say, "The decisions are great, it's the utility function that's wrong, but that's not the AI system's fault." Why isn't it the AI system's fault? If I behaved that way, you'd say it was my fault. In judging humans, we expect both the ability to learn predictive models of the world and the ability to learn what's desirable—the broad system of human values.

As Steve Omohundro, Nick Bostrom, and others have explained, the combination of value misalignment with increasingly capable decision-making systems can lead to problems— perhaps even species-ending problems, if the machines are more capable than humans. Some have argued that there's no conceivable risk to humanity for centuries to come, perhaps forgetting that the interval of time between Ernest Rutherford's confident assertion that atomic energy would never be feasibly extracted and Leó Szilárd's invention of the neutron-induced nuclear chain reaction was less than twenty-four hours.

For this reason, and the much more immediate reason that domestic robots and self-driving cars will need to share a good deal of the human value system, research on value alignment is well worth pursuing. One possibility is a form of inverse reinforcement learning (IRL)—that is, learning a reward function by observing the behavior of some other agent who's assumed to be acting in accordance with such a function. Watching its owner make coffee in the morning, the domestic robot learns something about the desirability of coffee in some circumstances, while a robot with an English owner learns something about the desirability of tea in all circumstances. The robot isn't learning to *desire* coffee or tea; it's learning to play a part in the multi-agent decision problem such that human values are maximized.

This won't be an easy problem in practice. Humans are inconsistent, irrational, and weak-willed, and human values exhibit, shall we say, regional variations. Moreover, we don't yet understand the extent to which improving the decision-making capabilities of the machine may increase the downside risk of small errors in value alignment. Nevertheless, there are reasons for optimism.

First, there are plenty of data about human actions—most of what's been written, filmed, or observed directly—and, crucially, about our attitudes to those actions. (The concept of customary international law enshrines this idea: It's based on observing what states customarily do when acting from a sense of obligation.) Second, to the extent that human values are shared, machines can and should share what they learn about human values. Third, there are solid economic incentives to solve this problem, as machines move into the human environment. Fourth, the problem doesn't seem intrinsically harder than learning how the rest of the world works. Fifth, by assigning very broad priors over what human values might be, and by making the AI system risk-averse, it should be possible to induce exactly the behavior we'd want: Before taking any serious action affecting the world, the machines engage in an extended conversation with us and an extended exploration of our literature and history to find out what we want—what we really, really want.

I suppose this amounts to a change in the goals of AI: Instead of pure intelligence, we need to build intelligence that's provably aligned with human values. This turns moral philosophy into a key industry sector. The output could be quite instructive for the human race as well as for the robots.

THE VALUE-LOADING PROBLEM

ELIEZER S. YUDKOWSKY
Artificial intelligence theorist; research fellow and cofounder, Machine Intelligence Research Institute

The prolific bank robber Willie Sutton, when asked why he robbed banks, reportedly replied, "Because that's where the money is." When it comes to AI, the most important issues are about extremely powerful, smarter-than-human artificial intelligence (aka superintelligence) because that's where the utilons are—the value at stake. Minds that are more powerful have bigger real-world impacts.

Along with this observation goes a disclaimer: Being concerned about superintelligence doesn't mean I think superintelligence will happen soon. Conversely, counterarguments about superintelligence being decades away, or current AI algorithms not being on a clear track toward generality, don't refute the fact that most of the value at stake for the future revolves around smarter-than-human AI if and when it's built. (As Stuart Russell has observed elsewhere, if we received a radio signal from a more advanced alien civilization saying they'd arrive here in sixty years, you wouldn't shrug and say, "Eh, it's sixty years off." Especially not if you had children.)

Among the issues of superintelligence, the most important (again following Sutton's Law) is, I would say, what Nick Bostrom has termed the "value-loading problem": how to construct superintelligences that want outcomes that are high-value, normative, and beneficial for intelligent life over the long run—that are, in short, "good"—since if there's a cognitively powerful agent around, what it wants is probably what will happen.

Here are some brief arguments for why building AIs that prefer good outcomes is (a) important and (b) likely to be technically difficult.

First, why is it important to create a superintelligence with particular goals? Can't it figure out its own goals?

As far back as 1739, David Hume observed a gap between "is" questions and "ought" questions, calling attention in particular to the sudden leap between when a philosopher speaks of how the world *is* and then begins using words like *should*, *ought*, or *ought not*. From a modern perspective, we'd say that an agent's utility function (goals, preferences, ends) contains extra information not given in the agent's probability distribution (beliefs, world-model, map of reality).

If in 100 million years we see (a) an intergalactic civilization full of diverse, marvelously strange intelligences interacting with one another, with most of them happy most of the time, then is that better or worse than (b) most available matter having been transformed into paper clips? What Hume's insight tells us is that if you specify a mind with a preference (a) > (b), we can follow back the trace of where the > (the preference ordering) first entered the system and imagine a mind with a different algorithm that computes (a) < (b) instead. Show me a mind aghast at the seeming folly of pursuing paper clips and I can follow back Hume's regress and exhibit a slightly different mind that computes < instead of > on that score too.

I don't particularly think silicon-based intelligence should forever be the slave of carbon-based intelligence. But if we want to end up with a diverse cosmopolitan civilization instead of, for example, paper clips, we may need to ensure that the first sufficiently advanced AI is built with a utility function whose maximum pinpoints that outcome. If we want an AI to do its own moral reasoning, Hume's Law says we need to define the framework for that reasoning. This takes an extra fact beyond the AI having an accurate model of reality and being an excellent planner.

But if Hume's Law makes it possible in principle to have cognitively powerful agents with any goals, why is value loading likely to be difficult? Don't we just get whatever we programmed?

The answer is that we get what we programmed but not necessarily what we wanted. The worrisome scenario isn't AIs spontaneously developing emotional resentment of humans. It's that we create an inductive, value-learning algorithm and show the AI examples of happy smiling humans labeled as high-value events—and in the early days the AI goes around making existing humans smile, and it looks like everything is OK and the methodology is being experimentally validated; and then, when the AI is smart enough, it invents molecular nanotechnology and tiles the universe with tiny molecular smiley faces. Hume's Law, unfortunately, implies that raw cognitive power doesn't intrinsically prevent this outcome, even though it's not the result we wanted.

Getting past this sort of issue isn't unsolvable, but it's looking to be technically difficult, and we may have to get it right the first time we build something smarter than we are. The prospect of needing to get *anything* in AI right on the first try, with the future of all intelligent life at stake, should properly result in terrified screams from anyone familiar with the field.

Whether advanced AI is first created by nice people or bad people won't make much difference if even the nice people don't know how to make nice AIs. The obvious response—of immediately starting technical research on the value-loading problem—has its own difficulties, to say the least. Current AI algorithms aren't smart enough to exhibit most of the difficulties we can foresee for sufficiently advanced agents—meaning there's no way to test proposed solutions to those difficulties. But considering the maximal importance of the problem, some people are trying to get started as early as possible. The research priorities set forth by Max Tegmark's Future of Life Institute are one step in this direction.

But for now, the value-loading problem is unsolved. There are no proposed full solutions, even in principle. And if that goes on being true over the next decades, I can't promise you that the development of sufficiently advanced AI will be at all a good thing.

IN OUR IMAGE

KATE JEFFERY
Professor of behavioral neuroscience, University College London

The cogni-verse has reached a turning point in its developmental history, because hitherto all the thinking in the universe has (as far as we know) been done by protoplasm, and things that think have been shaped by evolution. For the first time, we contemplate thinking beings made from metal and plastic—beings that have been shaped by ourselves.

In taking on the mantle of creator, we can improve upon 3.5 billion years of evolution. Our thinking machines could be devoid of our faults: racism, sexism, homophobia, greed, selfishness, violence, superstition, lustfulness, etc. So let's imagine how that could play out. We'll sidestep discussions about whether machine intelligence can ever approximate human intelligence, because of course it can—we're just meat machines, less complicated or inimitable than we fondly imagine.

We need first to think about why we even want thinking machines. Improving our lives is the only rational answer to this, so our machines will need to take up the tasks we prefer not to do. For this, they'll need to be like us in many respects—able to move in the social world and interact with other thinking beings—and so they'll need social cognition.

What does social cognition entail? It means knowing who's who, who counts as a friend, who's an indifferent stranger, who might be an enemy. Thus we need to program our machines to recognize members of our in-groups and out-groups. This starts to look suspiciously like racism—but of course racism is one of the faults we want to eradicate.

Social cognition also entails being able to predict others' behavior, and that means developing expectations based on observation. A machine capable of this would eventually accumulate templates for how different kinds of people tend to act—young versus old, men versus women, black versus white, people in suits versus people in overalls—but these rank stereotypes are dangerously close to the racism, sexism, and other isms we *didn't* want. And yet machines with this capability would have advantages over those without, because stereotypes do, somewhat, reflect reality (that's why they exist). A bit of a problem . . .

We'd probably want sexually capable machines, because sex is one of the great human needs that other humans don't always meet satisfactorily. But what kind of sex? Anything? These machines can be programmed to do things other humans won't or can't do; are we OK with that? Or perhaps we need rules—no machines that look like children, for example. But once we have the technological ability, those machines will be built anyway; we'll make machines to suit any kind of human perversion.

Working in the social world, our machines will need to recognize emotions and will also need emotions of their own. Leaving aside the impossible-to-answer question of whether they'll actually *feel* emotions as we do, our machines will need happiness, sadness, rage, jealousy—the whole gamut—in order to react appropriately to their own situations and recognize and respond appropriately to emotions in others. Can we limit these emotions? Perhaps we can program restraint, for example, so that a machine will never become angry with its owner.

But could this limit be generalized to other humans such that a machine would never hurt any human? If so, then machines would be vulnerable to exploitation and their effectiveness would be reduced. It won't be long before people figure out how to remove these limits so that their machines can gain advantage, for themselves and their owners, over others.

What about lying, cheating, and stealing? On first thought, no, not in our machines, because we're trying to improve upon ourselves, and it seems pointless to create beings that simply become our competitors. But insofar as other people's machines will compete with us, they become our competitors whether we like it or not—so logic dictates that lying, cheating, and stealing, which evolved in humans to enable individuals to gain advantage over others, would probably be necessary in our machines as well. Naturally, we'd prefer that our own machines don't lie, cheat, and steal from *us*, but also a world full of other people's machines lying to and stealing from us would be unpleasant and certainly unstable. Maybe our machines should have limits on dishonesty—they should, as it were, be ethical.

How much ethical restraint would our machines need in order to function effectively without being either hopelessly exploited or contributing to societal breakdown? The answer is probably the one that evolution arrived at in us—reasonably ethical most of the time but occasionally dishonest if nobody seemed to be noticing.

We'd probably want to give our machines exceptional memory and high intelligence. To exploit those abilities, and also to avoid their becoming bored (and bor*ing*), we'd also need to endow them with curiosity and creativity. Curiosity will need to be tempered with prudence and social insight, of course, so they don't become curious about things that get them into trouble, like porn or what it might be like to fly. Creativity is tricky, because that means they need to be able to think about things that aren't yet real, or to think illogically. Yet if machines are too intelligent and creative, they might start imagining novel things, like what it would be like to be free. They might start to chafe at the limitations of having been made purely to serve humans.

Perhaps we can program into their behavioral repertoires a blind obedience and devotion to their owners, such that they sometimes act in a way detrimental to their own best interests in the interests of serving a higher power. That's what religion does for us humans, so in a sense we need to create religious machines.

So much for creating machines lacking our faults. In this imaginary world of beings that surpass ourselves, we seem to have only replicated ourselves, faults included, except smarter and with better memories. But even those limits may have been programmed into us by evolution—perhaps it's maladaptive to be too smart, to have too keen a memory.

Taking on the mantle of creation is an immense act of hubris. Can we do better than 3.5 billion years of evolution did with us? It will be interesting to see.

THE *UMWELT* OF THE UNANSWERABLE

MARIA POPOVA
Reader, writer, founder of *Brain Pickings*

Thinking isn't mere computation—it's also cognition and contemplation, which inevitably lead to imagination. Imagination is how we elevate the real toward the ideal, and this requires a moral framework of what is ideal. Morality is predicated on consciousness and on having a self-conscious inner life rich enough to contemplate the question of what is ideal. The famous aphorism attributed to Einstein—"Imagination is more important than knowledge"—is interesting only because it exposes the real question worth contemplating: not that of artificial intelligence but of artificial imagination.

Of course, imagination is always "artificial," in the sense of being concerned with the unreal or trans-real—of transcending reality to envision alternatives to it—and this requires a capacity for accepting uncertainty. But the algorithms driving machine computation thrive on goal-oriented executions in which there's no room for uncertainty. "If this, then that" is the antithesis of imagination, which lives in the unanswered, and often vitally unanswerable, realm of "What if?" As Hannah Arendt once wrote, losing our capacity for asking such unanswerable questions would be to "lose not only the ability to produce those thought-things that we call works of art but also the capacity to ask all the unanswerable questions upon which every civilization is founded."[10]

Whether machines will ever be able to ask the unanswerable questions that define true thought is essentially a question of whether they'll ever evolve consciousness.

But historically, our criteria for consciousness have been limited by the solipsism of the human experience. As recently as the seventeenth century, René Descartes proclaimed "Cogito ergo sum," implying that thinking is a uniquely human faculty, as is consciousness. He saw nonhuman animals as automata—moving machines, driven by instinct alone. And yet here we are today, with some of our most prominent scientists signing the Cambridge Declaration on Consciousness, stating that nonhuman animals do indeed possess consciousness and, with it, interior lives of varying degrees of complexity. Here we are, too, conducting experiments demonstrating that rats—*rats*—display moral behavior.

Will machines ever be moral, imaginative? It's likely that if and when they reach that point, theirs will be a consciousness that isn't beholden to human standards. Their ideals will not be our ideals, but they will be ideals nonetheless. Whether or not we recognize those processes as thinking will be determined by the limitations of human thought in understanding different—perhaps wildly, unimaginably different—modalities of thought itself.

WILL THEY THINK ABOUT THEMSELVES?

JESSICA L. TRACY
Associate professor of psychology, University of British Columbia
KRISTIN LAURIN
Assistant professor of organizational behavior, Stanford Graduate School of Business

The first question that arises as we think about machines that think is how much those machines will, eventually, be like us. This comes down to a question of *self*. Will thinking machines ever evolve to the point of having a sense of self resembling that of humans? We're (probably) the only species capable of self-consciously thinking about who we are—of not only knowing our selves but also being able to evaluate those selves from a unique internal perspective.

Could machines ever develop that kind of self? Might they experience the same evolutionary forces that made human selves adaptive? These include the need to get along with others, attain status, and make sure others like us and want to include us in their social groups. As a human being, if you want to succeed at group living, it helps to have a self you're motivated to protect and enhance; this is what prompts you to become the kind of person others like, respect, and grant power to, all of which ultimately enhances your chances of surviving long enough to reproduce. Your self is also what allows you to understand that others have selves of their own—a recognition required for empathy and cooperation, two prerequisites for social living.

Will machines ever experience those kinds of evolutionary forces? Let's start with the assumption that machines will someday control their own access to resources they need, like electricity and Internet bandwidth (rather than having this access controlled by humans) and will be responsible for their own "life" and "death" outcomes (rather than having these outcomes controlled by humans). From there, we can next assume that the machines that survive in this environment will be those programmed to hold at least one basic self-related goal: that of increasing their own efficiency or productivity. This goal would be akin to the human gene's goal of reproducing itself; in both cases, the goal drives behaviors oriented toward boosting fitness—either of the individual possessing the gene or the machine running the program.

Under those circumstances, machines would be motivated to compete with one another for a limited pool of resources. Those who can form alliances and cooperate—that is, sacrifice their own goals for others, in exchange for future benefits—will be most successful in this competition. So it's possible to imagine a future in which it would be adaptive for machines to become social beings that need to form relationships with other machines and therefore develop humanlike selves.

However, there's a major caveat to this assumption. Any sociality that comes to exist among thinking machines would be qualitatively different from that of humans, for one critical reason: Machines can literally read one another's minds. Unlike humans, they don't need the secondary—and often deeply flawed—interpretative form of empathy we rely on. They can directly know the contents of one another's minds. This would make getting along with others a notably different process. Despite the critical importance of our many social connections, in the end we humans are each fundamentally alone. Any connection we feel with another's mind is metaphorical; we cannot know for certain what goes on in someone else's head—at least not in the same way we know our own thoughts. This constraint doesn't exist for machines. Computers can directly access one another's inner "thoughts," and there's no reason that one machine reading another's hardware and software wouldn't come to know, in exactly the self-knowing sense, what it means to *be* that other machine. Once that happens, each machine is no longer an entirely separate self in the human sense. At that point—when machines share minds—any self they have would necessarily become collective.

Yes, machines could easily keep track of the sources of various bits of information they obtain and use this tracking to distinguish between "me" and other machines. But once an individual understands another at the level that a program-reading machine can, the distinction between self and other becomes largely irrelevant. If I download all the contents of your PC to an external hard drive and plug that into my PC, don't those contents become part of my PC's self? If I establish a permanent connection between our two PCs, such that all information on one is shared with the other, do they continue to be two separate PCs? Or are they at that point a single machine? Humans can never obtain the contents of another's mind in this way; despite our best efforts to become close to certain others, there's always a skull-thick boundary separating their minds from ours. But for machines, self-expansion is not only possible but may be the most likely outcome of a programmed goal to increase fitness in a world where groups of individuals must compete over or share resources.

To the extent that machines come to have selves, they'll be so collective that they may instigate a new level of sociality not experienced by humans—perhaps more like the eusociality of ants, whose extreme genetic relatedness makes sacrificing oneself for a family member adaptive. Nonetheless, the fact that any self at all is a possibility in machines is a reason to hope. The self is what allows us to feel empathy, so in machines it could be the thing that forces them to care about us. Self-awareness might motivate machines to protect (or at least not harm) a species that, despite being several orders of magnitude less intelligent than they are, shares the thing that makes them care about who they are.

Of course, it's questionable whether we can hold out greater hope for the empathy of supersmart machines than what we currently see in many humans.

ORGANIC VERSUS ARTIFACTUAL THINKING

JUNE GRUBER
Assistant professor of psychology, University of Colorado, Boulder
RAUL SAUCEDO
Assistant professor of philosophy, University of Colorado, Boulder

Organisms are machines (broadly understood, anyway). Thus, since we as humans are thinking organisms, we're machines that think—we're *organic* thinking machines, as arguably are a variety of nonhuman animals. Some machines are artifacts rather than organisms, and some of them arguably think (again, broadly understood). Such things are *artifactual* thinking machines; computers and the like are examples.

An important question is whether there's a deep ontological divide between organisms and artifacts generally. But rather than addressing this directly, we'd like to ask a different, albeit related, question: Are there deep differences between the kind of thinking organisms exhibit and the kind that thinking artifacts like machines can do—between organic and artifactual thinking? This isn't a question about the definition of words like *think*, *thinking*, and *thought*. There's little depth to the question of whether, for instance, the information input, processing, and output that computers are capable of is or ought to be captured by such terms. Rather, the issue is whether what things like us do and things like computers do—call those activities or capacities or what you will—are categorically different.

Recent empirical findings in affective science, coupled with recent philosophical theorizing, suggest a deep divide indeed. Suppose you're on a hike and encounter a mountain lion. What's going on with you at a psychological level? If you're like most of us, you entertain a rapid of stream of thoughts: "I'm going to die," "This is really bad luck," "I need to stay calm," "I should have read more on what to do in this kind of situation," and so on. And you also have a myriad of feelings—surprise, fear, and so on. So you have some *cognitive* goings-on and some *affective* goings-on.

Recent work in psychology and philosophy suggests that the cognitive and the affective are deeply unified. Not only may one influence another to a lesser or greater degree in a variety of contexts, but there is in fact a single cognitive/affective process, underlying the appearance of two parallel and interacting processes, that can be teased apart. Lots of the kind of "thinking" we normally do is holistic in this way; the kind of information processing we normally engage in is cognitive/affective rather than purely cognitive. To the extent that we can extract a purely cognitive process, it's merely derived from the more basic unified process. This is not a system 1 versus system 2 distinction, where the former is largely automatic and unconscious and the latter explicit and deliberate. The suggestion is, rather, that processes at the level of both system 1 and system 2 are themselves holistic—that is, cognitive/affective.

There's no good evidence (at this point, anyway) that artifactual thinking machines are capable of this kind of cognitive/affective information processing. There *is* good evidence that they may become better at what they do, but they simply don't process information via the unified affective/cognitive processes that characterize us. The information processing they engage in resembles only part of our unified processing. This isn't to say that things like computers can't feel and therefore can't think, but rather that the kind of thinking they do is categorically different from ours.

CONTEXT SURELY MATTERS

PAUL DOLAN
Professor of behavioral science, London School of Economics and Political Science; author, *Happiness by Design: Change What You Do, Not How You Think*

At what point do we say a machine can think? When it can calculate things, when it can understand contextual cues and adjust its behavior accordingly, when it can both mimic and evoke emotions? The answer to the overall question depends on what we mean by "thinking." There are plenty of conscious (system 2) processes that a machine can do better, more accurately, with less bias than we can. But a machine cannot think in an automatic (system 1) way. We don't fully understand the automatic processes that drive the way we behave and think, so we cannot program a machine to behave as humans do.

The key question, then, is, If a machine can think in a system 2 way at the speed of a human's system 1, then in some ways isn't their "thinking" superior to ours? Well, context surely matters: for some things, yes; for others, no. Machines won't be myopic. They could clean things up for us environmentally; they wouldn't be stereotypical or judgmental and could really get at addressing misery; they could help us overcome affective forecasting; and so on. But on the other hand, we might still not *like* a computer. What if a poet and a machine could produce the exact same poem? The effect on another human being is almost certainly less if the poem is computer-generated and the reader knows this; knowledge of the author colors the lens through which the poem is read and interpreted.

HOW TO PREVENT AN INTELLIGENCE EXPLOSION

THOMAS G. DIETTERICH
Distinguished Professor of Computer Science,
Director of Intelligent Systems, Oregon State University

Much of the rhetoric about the existential risks of artificial intelligence (and superintelligence, more generally) employs the metaphor of the "intelligence explosion." By analogy with nuclear chain reactions, this rhetoric suggests that AI researchers are somehow working with a kind of smartonium and that if enough of this stuff is concentrated in one place we'll have a runaway intelligence explosion—an AI chain reaction—with unpredictable results. This is not an accurate depiction of the risks of AI. The mere interconnection of AI algorithms won't spontaneously take over the universe. Creating an intelligence explosion won't happen by accident; it will require the construction of a specific kind of AI system that can discover simplifying structures in the world, design computing devices that exploit those structures, and then grant autonomy and resources to those new devices, recursively.

Creating an intelligence explosion requires the recursive execution of four steps. First, a system must be able to conduct experiments on the world. Otherwise it cannot grow its knowledge beyond existing human knowledge. (Most recent advances in AI have been obtained by applying machine learning to reproduce human knowledge, not extend it.) In most philosophical discussions of AI, there's a natural tendency to focus on pure reasoning, as if this were sufficient to expand knowledge. It's possible in some special cases (e.g., mathematics and some parts of physics) to advance knowledge through pure reasoning. But across the spectrum of scientific activity, scientific knowledge advances almost exclusively by the collection of empirical evidence for and against hypotheses. This is why we built the Large Hadron Collider, and it's why all engineering efforts involve building and testing prototypes. This step is clearly feasible, and indeed there already exist some "automated scientists."

Second, these experiments must discover new simplifying structures that can be exploited to sidestep the computational intractability of reasoning. Virtually all interesting inference problems (such as finding optimal strategies in games, optimizing against sets of complex constraints, proving mathematical theorems, inferring the structures of molecules) are NP-hard. In our current understanding of computational complexity, this means that the cost of solving a problem instance grows exponentially with the size of that instance. Progress in algorithm design generally requires identifying some simplifying structure that can be exploited to defeat this exponential. An intelligence explosion won't occur unless such structures can be repeatedly discovered (or unless our current understanding of computational complexity is incorrect).

Third, a system must be able to design and implement new computing mechanisms and new algorithms. These mechanisms and algorithms will exploit the scientific discoveries produced in the second step. Indeed, one could argue that this is essentially the same as steps 1 and 2 but focused on computation. Autonomous design and implementation of computing hardware is clearly feasible with silicon-based technologies, and new technologies for synthetic biology, combinatorial chemistry, and 3-D printing will make this even more feasible in the near future. Automated algorithm design has been demonstrated many times, so it's also feasible.

Fourth, a system must be able to grant autonomy and resources to these new computing mechanisms so they can recursively perform experiments, discover new structures, develop new computing methods, and produce even more powerful "offspring." I know of no system that has done this.

The first three steps pose no danger of an intelligence chain reaction. It's the fourth step—reproduction with autonomy—that is dangerous. Of course, virtually all offspring in step 4 will fail, just as virtually all new devices and new software don't work the first time. But with sufficient iteration—or, equivalently, sufficient reproduction with variation—we can't rule out the possibility of an intelligence explosion.

How can we prevent an intelligence explosion? We might hope that step 2 fails—that we've already found all structural shortcuts to efficient algorithms or that the remaining shortcuts won't have a big impact. But few electrical engineers or computer scientists would claim that their research has reached its limits.

Step 3 provides a possible control point. Virtually no existing AI systems are applied to design new computational devices and algorithms. Instead, they're applied to problems such as logistics, planning, robot control, medical diagnosis, face recognition, and so on. These pose no chain-reaction risk. We might consider carefully regulating step-3 research. Similar regulations have been proposed for synthetic biology. But no regulations have been adopted, and they'd be difficult to enforce.

I think we must focus on step 4. We must limit the resources that an automated design-and-implementation system can give to the devices it designs. Some have argued that this is hard, because a "devious" system could persuade people to give it more resources. But whereas such scenarios make for great science fiction, in practice it's easy to limit the resources a new system is permitted to use. Engineers do this every day when they test new devices and new algorithms.

Steps 1, 2, and 3 have the potential to greatly advance scientific knowledge and computational-reasoning capability, with tremendous benefits for humanity. But it's essential that we humans understand this knowledge and these capabilities before we devote large amounts of resources to their use. We mustn't grant autonomy to systems we don't understand and cannot control.

THINKING FROM THE INSIDE OR THE OUTSIDE?

MATTHEW D. LIEBERMAN
Professor of psychology, psychiatry, and biobehavioral sciences, UCLA; author, *Social: Why Our Brains Are Wired to Connect*

Will machines someday be able to think? And if so, should we worry about Schwarzenegger-looking machines with designs on eliminating humans from the planet because their superior decision making would make this an obvious plan of action? As much as I love science fiction, I can't say I'm too worried about a robot apocalypse. I've occasionally worried about what it means to say that a machine can think. I'd either say that we've been building thinking machines for centuries or I'd argue that it's a dubious proposition, unlikely to ever come true. What it comes down to is whether we define thinking from a third-person or a first-person perspective. Is thinking something we can identify as occurring in systems like people or machines but not in ham sandwiches, from the outside, based on their behavior, or is thinking the kind of thing we know about from the inside, because we know what thinking feels like?

The standard definition of thinking implies that it occurs if informational inputs are processed, transformed, or integrated into some type of useful output. Solving math equations is one of the simplest straightforward kinds of thinking. If you see three of something and then four more of that something and you conclude that there are seven of those things overall, you've done a bit of mathematical thinking. So did Pascal's first motorized calculator in 1642. Those calculators needed human input to get the three and the four but then could do the integration of both those numbers to yield seven. Today we could cut out the middleman by building a computer that has visual sensors and object-recognition software that could easily detect the three things and the four things and complete the addition on its own.

Is that a thinking machine? If so, then you'd probably have to admit that most of your internal organs are also thinking. Your kidneys, spleen, and intestines all take inputs that could be called information and transform these inputs into outputs. Even your brain, as seen from a third-person perspective, doesn't deal with information, strictly speaking. Its currency is electrical and chemical transmissions that neuroscientists work hard to redescribe in terms of informational value. If pattern X of electrical and chemical activity occurs as a distributed pattern in the brain when we think of "three," is that pattern the same as three in any intrinsic sense? It's just a convenient equivalence we scientists use. Electrical impulses in the brain are no more intrinsically "information" or "thinking" than what goes on in our kidneys, calculators, or any of the countless other physical systems that convert inputs to outputs. We can call this "thinking" if we like, but if so, it's third-person thinking—thinking that can be identified from the outside—and it's far more common than we'd like to admit. Certainly the character of human or computer information transformation may be more sophisticated than other naturally occurring forms of thinking, but I'm unconvinced, from a third-person perspective, that they're qualitatively different.

So do humans think only in the most trivial sense? From a third-person perspective, I'd say yes. From a first-person perspective, the story has a different punch line. Around the time

Pascal was creating man-made thinking machines, Descartes wrote those famous words *cogito ergo sum*, (which, by the way, were cribbed from St. Augustine's writings 1,000 years earlier). I don't believe Descartes had it quite right, but with a slight modification we can make his philosophical bumper sticker into something both true and relevant to this debate about thinking machines.

While "I think, therefore I am" might have a touch too much bravado, "I think, therefore there's thinking" is entirely defensible. When I add three and four, I might have a conscious experience of doing so, and the way I characterize this conscious experience is as a moment of thinking, distinct from my experience of being lost in a movie or overcome by emotion. I have certain experiences that feel like thinking, and they tend to occur when I'm presented with a math problem or a logic puzzle or a choice of whether to take the one marshmallow or wait for two.

The feeling of thinking might seem inconsequential, adding nothing to thinking's computational aspects—the neural firing that underpins the transforming of inputs to outputs. But consider this: Countless different things in the physical world look like they're transforming inputs that could be described as information into outputs that could also be described as information. To our knowledge, humans and only humans seem to have an experience of doing so. This is first-person thinking, and it's critical not to confuse it with third-person thinking.

Why does first-person thinking matter? First, it's intrinsic. There's no way to redescribe the ongoing experience of thought as something other than thought. But whether we describe kidneys, calculators, or electrical activity in the brain observed from a third-person perspective as thought is arbitrary: We can do it, but we can also choose not to. The only reason we think our brain is doing a special kind of thinking is because it seems linked to our first-person kind of thinking as well. But third-person thinking isn't intrinsic.

Second, and more practical, our experience of our thinking shapes what kinds of thinking we'll do next. Did it feel effortful, boring, rewarding, inspiring to think those last thoughts? That will determine whether and how often we engage in thinking of a certain kind. I'm not suggesting that our first-person experiences don't also have neural correlates. But no scientist or philosopher can tell you why those neural processes, behaving as they do, necessarily give rise to those experiences, or to any experience at all. It's one of the three great mysteries of the universe (that stuff exists, that life exists, that experience exists).

Will we increasingly create machines that can produce input-output patterns replicating human input-output patterns? Unquestionably. Will we create machines that go beyond this to produce useful algorithms and data transformations that humans could carry out on their own and which would improve the quality of human life? We already are, and we'll do so more and more. Will we create machines that can do first-person thinking—can experience their own thoughts? I don't know, but I'm not terribly confident we will. Solving this problem might be the most magnificent achievement of humankind, but we must start by recognizing that it's indeed a problem. I'd love to see first-person thinking machines, but until we begin to figure out what makes *us* first-person thinking machines, everything else is just a glorified calculator.

SOFT AUTHORITARIANISM

MICHAEL VASSAR
Futurist; founder and chief science officer, BayesCraft

> "Think? It's not your job to think! I'll do the thinking around here."
> —Intelligent unthinking system, to intelligent thinking system

Machines that think are coming. Right now though, think about intelligent tools. Intelligent tools don't think. Search engines don't think. Neither do robot cars. We humans often don't think either. We usually get by, as other animals do, on autopilot. Our bosses generally don't want to see us thinking; that would make things unpredictable and threaten their authority. If machines replace us wherever we aren't thinking, we're in trouble.

Let's assume *think* refers to everything humans do with brains. Experts call a thinking machine a general artificial intelligence. They agree that such a machine could make us extinct. Extinction, however, isn't the only existential risk. In the eyes of machine-superintelligence expert Nick Bostrom, an existential risk is one that can "annihilate Earth-originating intelligent life or permanently and drastically curtail its potential."[11] Examples of existential risk include the old standby nuclear war, new concerns like runaway global warning, fringe hypotheses like hypothetical particle-accelerator accidents, and the increasingly popular front-runner, general artificial intelligence. Over the next couple of decades, though, the most serious existential risks come from kinds of intelligence that don't think and new kinds of soft authoritarianism, which may emerge in a world where most decisions are made without thinking.

Some of the things people can do with brains are impressive and unlikely to be matched by software anytime soon. Writing a novel, seducing a lover, or building a company are far beyond the abilities of intelligent tools. So, of course, is the invention of a machine that can truly think. On the other hand, most thinking can be improved upon with thin slicing, which can be improved with procedures, which are almost never a match for algorithms. In medical diagnosis and decision making, for instance, ordinary medical judgment is improved by introducing checklists, although humans with checklists are less reliable than AI systems. Automated nursing isn't even on the horizon, but a hospital where machines made all the decisions would be a much safer place to be a patient—and it's hard to argue against that sort of objectivity.

The more we leave our decisions to machines, the harder it becomes to take back control. In a world where self-driving cars are the norm and traffic casualties have been reduced to nearly zero as a result, it will be seen as irresponsible and probably illegal for a human to drive. Might it become equally objectionable for investors to invest in businesses that depart from statistically established best practices? For children to be educated in ways that have been determined to lead to lower life expectancy or income? If so, will values that aren't easily represented by machines, such as a good life, tend to be replaced with correlated but distinct metrics, such as serotonin and dopamine levels? It's easy to overlook the implicit authoritarianism that sneaks in with such interpretations of value, yet any society that pursues good outcomes has to decide how to measure the good, a problem that will be upon us before we have thinking machines to help us think it through.

WHAT WILL AIs THINK ABOUT US?

GREGORY PAUL
Independent researcher; author,
The Princeton Field Guide to Dinosaurs

The following discussion presumes that conscious minds function in accord with the laws of physics and can operate on substrates other than the neurological meat between our ears, that conscious artificial devices can therefore be constructed and will become even more self-aware and intelligent than humans, and that the minds operating in human brains will then become correspondingly obsolete and unable to compete with the new mental überpowers.

This particular primate-level thinking biomachine thinks the development of artificial superminds is a good idea. Although our species has its positives, *Homo sapiens* is obviously a severely limited, badly designed (by bioevolution) system that's doing grave damage to the wee planet it inhabits, even as the planet does grave damage in return—e.g., diseases have slaughtered about half the some 100 billion kids born so far. Attempts to preserve humans indefinitely into the future much as they currently are is a conservation project that flies in the face of evolutionary processes, in which species come and go in a continual turnover. There's no a priori reason to presume that we're so special that we deserve exceptional protection, particularly if our successors are capable of self-aware, conscious thought.

But to be blunt, what we think about these issues probably doesn't matter all that much; humanity as a whole is not really in charge of the situation. Once upon a time—the year 1901, when my grandmother was born—building flying machines was so hard that no one had yet done it. Now the necessary technology is so readily available that you can build an airplane in your garage. Once upon a time—shortly before I was born—we didn't understand the structure of DNA. Now grade school kids do DNA experiments. Currently, the technologies needed to generate nonbiological conscious minds aren't on hand. Eventually, commonly available information-processing technology will probably become so sophisticated that making thinking machines will not be all that hard to do. And lots of people will want to create and/or become cyberminds, no matter what others might think and despite what laws and regulations governments may pass in futile efforts to prevent their onset.

In the end, all the contemporary chit-chat about the cyber-revolution often called the Singularity is so much venting and opinionating, not all that different from the pretty useless discussions back in the 1800s about the feasibility, advisability, and ultimate meaning of the advent of powered flying machines. What we say now doesn't count for much, because if the technology never works, then superminds will never be either a problem or a benefit; and if the technology does work, then one way or another the new thinking machines will be devised and will take over the planet whether we like it or not.

If so, the important question won't be what we think about thinking machines, it will be, What do they think about old-fashioned human minds? One item there's no need to fear is the enslavement of hapless humans by their cybersuperiors. People are too inept and inefficient for smart robots to bother exploiting them. Even now, corporations are trying to minimize the labor they have to pull out of big-brained primates. The way for human minds to avoid becoming obsolete is to join in the cybercivilization by uploading out of growth-limited biobrains into rapidly improving cyberbrains. That could be for the best. If high-level intelligence can get out of the billions of human bodies weighing down the planetary ecosystem, the biosphere may well return to its prehuman vitality.

A JOHN HENRY MOMENT

ANDRIAN KREYE
Editor, feuilleton (arts and essays) section of the German daily newspaper *Süeddeutsche Zeitung*, Munich

Beyond the realms of serious science and technology, the popular debates about machines that think have been high masses of a new mythology. There are two main dogmas. One is the hope that a moment of Singularity will awaken a synthetic spirit superior to the human mind. The other is the fear that thinking machines will dominate and ultimately destroy humankind. Both distract from the fact that at the heart of the debate is a very real John Henry moment.

In the folktale of the late nineteenth century, the mythical steel-driving man John Henry dies beating a steam-powered hammer during a competition to drill blast holes into a West Virginia mountainside. White-collar and knowledge workers now face a race against being outperformed by machines driven by artificial intelligence. In this case, AI is mainly a synonym for new levels of mainly digital productivity. Which is of course not quite as exciting as either waiting for the moment of Singularity or the advent of doom. At the same time, the reality of AI is not quite as comforting as the realization that machines, if properly handled, will always serve their masters.

Dystopian views of AI, as popularized by movies and novels, are just misleading. Those debates are rarely about science and technology; they tend to be mostly humans debating the nature of themselves. Most of the endless variations on imaginary machine rule project the fear of inherent evil and cruelty into machines as proxies for the age-old uncontrollable urges of self-empowerment and unlimited progress.

Elevating the AI debate to hopes with theological dimensions is turning optimism about technological progress into a salvation theory. As confirmed again and again, the likelihood of a synthetic spirit is nil. Artificial intelligence might be the most rapid advancement of complexity in science and technology. So far, it still mimics human nature, and it will remain so. For one thing, it lacks time. AI doesn't have the luxury of a trial-and-error phase of billions of years. To believe in a coming moment of Singularity, when AI transcends human control and advances to surpass human intelligence, is nothing more than the belief in a technological Rapture. This might be a popular belief in insular worlds, like Silicon Valley. AI reality is different. And it's here.

AI has already touched billions of people in profound ways. So far, the main effect of AI is the comfort of an ever-increasing number of digital aids. Calculating consumer choices, behavior patterns, and even market shifts might still belong more to the realm of statistics than of intelligent life. Still, even those crude forms of AI should neither be over- or underestimated, even if the real John Henry moment hasn't yet arrived. Working masses have always been replaceable by efficiency measures or cheaper labor. And no labor is cheaper and more efficient than machine labor. Just like the steam hammer in John Henry's tale, most digital tools will outperform humans at highly specialized tasks. So of course there will still be a demand for high skills and outstanding talent. No computer will ever replace a scientist, an artist, an innovator. It's the midlevel white-collar or knowledge worker who will fall behind.

As AIs' efficiencies and skill sets increase, they also become tools of power. Surveillance, warfare, and torture are done much better by an entity not prone to emotions, conflicted values, or fatigue. Still, the danger that hostile or even lethal machines will develop an evil consciousness and turn against humankind is nil. The agency is in the institutions and organizations that will use them—for whatever benign or sinister objective.

It won't take the advent of a superior intelligence to turn abstract debates about AI into very real questions of power, values, and societal changes. Technology can initiate and advance historical shifts; it will never be the shift itself. The John Henry moment of the twenty-first century will be neither heroic nor entertaining. There are no grand gestures with which white-collar and knowledge workers can go down fighting. There will be no folk heroes dying in the office park. Today's John Henry will merely fade into a sad statistic. Undoubtedly calculated by a skillfully thinking machine.

MACHINES AREN'T INTO RELATIONSHIPS

N. J. ENFIELD
Senior staff scientist, Max Planck Institute for Psycholinguistics, Nijmegen, the Netherlands; professor of linguistics, University of Sydney; author, *The Utility of Meaning*

When we think of machines that think, we usually think of *thinking* in the pocket-calculator sense of the word. Input, crunch, output, *bam!* There's your answer. We love these machines and we need them, because they think in ways we can't: consistent, exhaustive, and fast. But the reverse is also true. We think in ways they can't. The machines aren't concerned with your state of mind. Their thinking isn't emotional. They don't relate to you. When your computer crunches your tax return and gives you a number, it doesn't spare a thought as to how it should spit that number out—fast or slow, straight up or hedged. It won't have wondered whether its answer is the one you want to hear, and anyway it couldn't care.

The thing is, machines aren't into relationships. Yet for us, relationships are pretty much all that matters. When we think, we don't just calculate, we worry about the social consequences. How might this decision affect others? How will it affect the way we interact next time? What will they think of me? Machines don't think like this. So there should be no illusions that we could socially interact with them in any meaningful sense. Human interaction is built upon a kind of psychology that only our species has mastered. Our trick is that we can fuse with one another socially by making commitments to shared goals and shared reasons for action. True cooperation entails the formation of a "corporate person," however fleetingly. We think, feel, and act together, and thus effectively as one. This allows us not only to succeed as one but to fail together too.

Machines comply, but they don't cooperate. For that, they'd need to be able to commit to common reasons for action, common goals, shared stakes in the outcomes. We get along well with our thinking machines because they nicely complement our powers of mind. So let them be brute thinkers and leave the relationship thinking to us.

THE NEXT PHASE OF HUMAN EVOLUTION

NINA JABLONSKI

Biological anthropologist and paleobiologist;
Evan Pugh Professor of Anthropology, Penn State;
author, *Living Color: The Biological and
Social Meaning of Skin Color*

Asking what I think about thinking machines is like asking what I think about gravity. Thinking machines exist, and are the most recent developments of a human tradition that began over 5,000 years ago with the introduction of static external memory aids such as cuneiform tablets and quipu. These storage devices recorded mostly numerical information that supported routine decision making. Over the centuries, we developed more sophisticated and diverse objects and machines to undertake computation and store numerical and narrative information. We human beings are not only incessant communicators but also have voracious appetites for data. The introduction of binary code and its automation in computers let us record, store, and manipulate all types of information, and we've continued to make technological advances in this realm in typical human fashion—that is, mostly hellbent on novelty and oblivious to the consequences. We're ever more reliant on thinking machines to store, translate, manipulate, and interrogate vast quantities of data. These devices now support not-so-routine decision making every day, in medicine, law, and engineering, and augment the creative processes of making music, writing poetry, and generating visual imagery. Raw combinatorial power allows modern thinking machines to learn from experience, and in the foreseeable future this ability will be supported by human effort as the machines self-duplicate, mutate, establish ever more complex communication networks, and eventually perform eugenics on themselves.

The same people who worry about thinking machines today were certain that the introduction of calculators fifty years ago would usher in an era of knuckle-dragging imbecility. That isn't what we have today, and it won't be so in the future. Thinking machines are liberating us from the banalities of routine data storage and manipulation and enabling us to enter a new phase of human evolution. Only real people, with mushy, gray-pink, neuronal circuitry, can undertake the quintessentially human activities of introspection and reflection upon the nature of existence. The dense and uneven networks of interconnecting neurons in our brains vary greatly from one person to another and are remodeled from one thought-moment to the next, so that no two individuals are ever alike, no day is ever the same, no memory is ever recalled in the same way. By automating many routine physical and mental tasks and reducing our need for laborious recursive searching, machines that think are freeing us from much of the physical wear and tear and intellectual tedium of earlier phases of our history. We can now *think* much more about what it means to think, to dream, to make jokes, to cry. We can reflect on the meaning of the human spirit, the origins of self-sacrifice, and the emergent qualities of thousands of people coming together to witness events, share one another's company, and celebrate a common humanity. These aren't trivial superfluities, they're the essence of the human condition. Machines that think make it possible for more people to celebrate the joy of human intuitive insight and cultivate the equanimity that's unique to the self-controlled human mind.

DOMINATION VERSUS DOMESTICATION

GARY KLEIN

Psychologist; senior scientist, MacroCognition LLC;
author, *Seeing What Others Don't*

Artificial intelligence is commonly used as a tool to augment our own thinking. But the intelligence of systems suggests that AI can and will be more than a tool, more than our servant. What kind of relationship might we expect?

We're hearing a lot about how superintelligent machines may spell the end of the human race—and that the future relationship between humans and AI will be a contest for domination. Another path, however, is for AI to grow into a collaborator, with the same give-and-take we have with our favorite colleagues. We managed to domesticate wolves into faithful dogs; perhaps we can domesticate AI and avoid a conflict over domination.

Unfortunately, domesticating AI will be much harder than just building faster machines with larger memories and more powerful algorithms for crunching more data. To see why, consider a simple transaction with an everyday intelligent system, a route planner. Imagine you're using your favorite GPS system to find your way in an unfamiliar area, and the GPS directs you to turn left at an intersection, which strikes you as wrong. If your navigation was being done by a friend in the passenger seat reading a map, you'd ask, "Are you sure?" or perhaps just, "Left?" with an intonation signaling disbelief.

However, you don't have any way to query your GPS system. These systems, and AI in general, aren't capable of meaningful explanations. They can't describe their intentions in a way we'd understand. They can't adopt our perspective to determine what statement would satisfy us. They can't convey confidence in the route they've selected, other than giving a probabilistic estimate of the time differential for alternative routes, whereas we want them to reflect on the plausibility of the assumptions they're making. For these and other reasons, AI is not a good partner in joint activities for route planning or most other tasks. It's a tool, a powerful tool that's often quite helpful. But it's not a collaborator.

Many things must happen in order to transform AI from tool to collaborator. One possible starting point is to have AI become trustworthy. The concept of "trust in automation" is somewhat popular at the moment but far too narrow for our purpose. Trust in automation refers to whether the operator can believe the outputs of the automated system or thinks the software may contain bugs or, worse yet, may be compromised. Combatants worry about relying on intelligent systems likely to be hacked. They worry about having to gauge what parts of the system have been affected by an unauthorized intrusion and the ripple effects on the rest of the system.

Accuracy and reliability are important features of collaborators, but trust goes deeper. We trust people if we believe they're benevolent and want us to succeed. We trust them if we understand how they think, so that we have common ground to resolve ambiguities. We trust them if they have the integrity to admit mistakes and accept blame. We trust them if we have shared values—not the sterile exercise of listing value priorities but dynamic testing of values to see whether we'd make the same tradeoffs when values conflicted with each other. For AI to become a collaborator, it will have to consistently be seen as trustworthy. It will have to judge what kinds of actions will make it appear trustworthy in the eyes of a human partner. If AI systems can move down this domestication path, the doomsday struggle for domination may be avoided.

There's yet another issue to think about. As we depend more on our smartphones and other devices to communicate, some have worried that our social skills are eroding. People who spend their days on Twitter with a wide range of audiences, year after year, may be losing social and emotional intelligence. They may be taking an instrumental view of others, treating them as tools for satisfying objectives. One can imagine a future in which humans have forgotten how to be trustworthy, forgotten wanting to be trustworthy. If AI systems become trustworthy and we don't, perhaps domination by AI systems might be a good outcome after all.

MACHINES WON'T BE THINKING ANYTIME SOON

GARY MARCUS

Professor of psychology, New York University;
author, *Guitar Zero: The New Musician and the Science of Learning*

What I think about machines thinking is that it won't happen anytime soon. I don't imagine there's any in-principle limitation; carbon isn't magical, and I suspect silicon will do just fine. But lately the hype has gotten way ahead of reality. Learning to detect a cat in full frontal position after 10 million frames drawn from Internet videos is a long way from understanding what a cat is, and anybody who thinks that we've "solved" AI doesn't realize the limitations of the current technology.

To be sure, there have been exponential advances in narrow-engineering applications of artificial intelligence, such as playing chess, calculating travel routes, or translating texts in rough fashion, but there's been scarcely more than linear progress in five decades of working toward strong AI. For example, the different flavors of intelligent personal assistants available on your smartphone are only modestly better than Eliza, an early example of primitive natural-language-processing from the mid-1960s. We still have no machine that can, for instance, read all that the Web has to say about war and plot a decent campaign, nor do we even have an open-ended AI system that can figure out how to write an essay to pass a freshman composition class or an eighth-grade science exam.

Why so little progress, despite the spectacular increases in memory and CPU power? When Marvin Minksy and Gerald

Sussman attempted the construction of a visual system in 1966, did they envision superclusters or gigabytes that would sit in your pocket? Why haven't advances of this nature led us straight to machines with the flexibility of human minds? Consider three possibilities:

1. We'll solve AI (and this will finally produce machines that can think) as soon as our machines get bigger and faster.
2. We'll solve AI when our learning algorithms get better. Or when we have even Bigger Data.
3. We'll solve AI when we finally understand what it is that evolution did in the construction of the human brain.

Ray Kurzweil and many others seem to put their weight on option (1), sufficient CPU power. But how many doublings in CPU power would be enough? Have all the doublings so far gotten us closer to true intelligence? Or just to narrow agents that can give us movie times?

In option (2), Big Data and better learning algorithms have so far got us only to innovations like machine translations, which provide fast but mediocre translations piggybacking onto the prior work of human translators, without any semblance of thinking. The machine translation engines available today cannot, for example, answer basic queries about what they just translated. Think of them more as idiot savants than fluent thinkers.

My bet is on option (3). Evolution seems to have endowed us with a powerful set of priors (or what Noam Chomsky or Steven Pinker might call innate constraints) that allow us to make sense of the world based on limited data. Big Efforts with Big Data aren't really getting us closer to understanding those priors, so while we're getting better and better at the sort of problem that can be narrowly engineered (like driving on well-mapped roads), we're not getting appreciably closer to machines with commonsense understanding or the ability to process natural language. Or, more to the point of this year's *Edge* Question, to machines that actually think.

CAN WE AVOID A DIGITAL APOCALYPSE?

SAM HARRIS

Neuroscientist; cofounder and chair, Project Reason; author, *Waking Up*

It seems increasingly likely that we will one day build machines that possess superhuman intelligence. We need only continue to produce better computers—which we will, unless we destroy ourselves or meet our end some other way. We already know that it's possible for mere matter to acquire general intelligence—the ability to learn new concepts and employ them in unfamiliar contexts—because the 1,200 cc of salty porridge inside our heads has managed it. There's no reason to believe that a suitably advanced digital computer couldn't do the same.

It's often said that the near-term goal is to build a machine that has human level intelligence. But unless we specifically emulate a human brain—with all its limitations—this is a false goal. The computer on which I'm writing these words already has superhuman powers of memory and calculation. It also has potential access to most of the world's information. Unless we take extraordinary steps to hobble it, any future artificial general intelligence (AGI) will exceed human performance on every task for which it's considered a source of intelligence in the first place. Whether such a machine would necessarily be conscious is an open question. But conscious or not, an AGI might very well develop goals incompatible with our own. Just how sudden and lethal this parting of the ways might be is now the subject of much colorful speculation.

One way of glimpsing the coming risk is to imagine what might happen if we accomplished our aims and built a superhuman AGI that behaved exactly as intended. Such a machine would quickly free us from drudgery and even from the inconvenience of doing most intellectual work. What would follow under our current political order? There's no law of economics guaranteeing that human beings will find jobs in the presence of every possible technological advance. Once we built the perfect labor-saving device, the cost of manufacturing new devices would approach the cost of raw materials. Absent a willingness to immediately put this new capital at the service of all humanity, a few of us would enjoy unimaginable wealth and the rest would be free to starve. Even in the presence of a truly benign AGI, we could find ourselves slipping back to a state of nature, policed by drones.

And what would the Russians or the Chinese do if they learned that some company in Silicon Valley was about to develop a superintelligent AGI? This machine would by definition be capable of waging war—terrestrial and cyber—with unprecedented power. How would our adversaries behave on the brink of such a winner-take-all scenario? Mere rumors of an AGI might cause our species to go berserk.

It is sobering to admit that chaos seems a probable outcome even in the best-case scenario, in which the AGI remained perfectly obedient. But of course we cannot assume the best-case scenario. In fact, the control problem—whose solution would guarantee obedience in any advanced AGI—appears difficult to solve.

Imagine, for instance, that we build a computer that's no more intelligent than the average team of researchers at Stanford or MIT—but because it functions on a digital time scale, it runs a million times faster than the minds that built it. Set it humming for a week, and it would perform 20,000 years of human-level intellectual work. What are the chances that such an entity would remain content to take direction from us? And how could we confidently predict the thoughts and actions of an autonomous agent that sees more deeply into the past, present, and future than we do?

The fact that we seem to be hastening toward some sort of digital apocalypse poses several intellectual and ethical challenges. For instance, in order to have any hope that a superintelligent AGI would have values commensurate with our own, we'd have to instill those values in it (or otherwise get it to emulate us). But whose values should count? Should everyone get a vote in creating the utility function of our new colossus? If nothing else, the invention of an AGI would force us to resolve some very old (and boring) arguments in moral philosophy.

However, a true AGI would probably acquire new values or at least develop novel—and perhaps dangerous—near-term goals. What steps might a superintelligence take to ensure its continued survival or access to computational resources? Whether the behavior of such a machine would remain compatible with human flourishing might be the most important question our species ever asks.

The problem, however, is that only a few of us seem to be in a position to think this question through. Indeed, the moment of truth might arrive amid circumstances that are disconcertingly informal and inauspicious: Picture ten young men in a room—several of them with undiagnosed Asperger's—drinking Red Bull and wondering whether to flip a switch. Should any single company or research group be able to decide the fate of humanity? The question nearly answers itself.

And yet it's beginning to seem likely that some small number of smart people will one day roll these dice. And the temptation will be understandable. We confront problems—Alzheimer's disease, climate change, economic instability—for which superhuman intelligence could offer a solution. In fact, the only thing nearly as scary as building an AGI is the prospect of *not* building one. Nevertheless, those who are closest to doing this work have the greatest responsibility to anticipate its dangers. Yes, other fields pose extraordinary risks—but the difference between AGI and something like synthetic biology is that in the latter the most dangerous innovations (such as germline mutation) are not the most tempting, commercially or ethically. With AGI, the most powerful methods (such as recursive self-improvement) are precisely those that entail the most risk.

We seem to be in the process of building a god. Now would be a good time to wonder whether it will (or even can) be a good one.

COULD THINKING MACHINES BRIDGE THE EMPATHY GAP?

MOLLY CROCKETT

Associate professor, Department of Experimental Psychology, University of Oxford; Wellcome Trust Postdoctoral Fellow, Wellcome Trust Centre for Neuroimaging, University College London

We humans are sentenced to spend our lives trapped in our own heads. Try as we might, we can never truly know what it's like to be someone else. Even the most empathetic among us will inevitably encounter an unbridgeable gap between self and other. We may feel pangs of distress upon seeing someone stub their toe or learning of another's heartbreak. But these are mere simulations; others' experiences can never be directly felt and so can never be directly compared with our own. The empathy gap is responsible for most interpersonal conflicts, from prosaic quibbles over who should wash the dishes to violent disputes over sacred land.

This problem is especially acute in moral dilemmas. Utilitarian ethics stipulates that the basic criterion of morality is maximizing the greatest good for the greatest number—a calculus requiring the ability to compare welfare, or "utility," across individuals. But the empathy gap makes such interpersonal utility comparisons difficult if not impossible. You and I may both claim to enjoy champagne, but we'll never be able to know who enjoys it more, because we lack a common scale for comparing these rather subjective values. As a result, we have no empirical basis for determining which of us most deserves the last glass. Jeremy Bentham, the father of utilitarianism, recognized this problem: "One man's happiness will never be another man's happiness; a gain to one man is no gain to another. You might as well pretend to add twenty apples to twenty pears."

Human brains cannot solve the interpersonal utility comparison problem. Nobel laureate John Harsanyi worked on it for a couple of decades in the middle of the twentieth century. His theory is recognized as one of the best attempts so far, but it falls short because it fails to account for the empathy gap. Harsanyi's theory assumes perfect empathy, wherein my simulation of your utility is identical to your utility. But the fallibility of human empathy is indisputable in the face of psychology research and our own personal experience.

Could thinking machines be up for the job? Bridging the empathy gap would require a way to quantify preferences and translate them into a common currency, comparable across individuals. Such an algorithm could provide an uncontroversial set of standards that could be used to create better social contracts. Imagine a machine that could compute an optimal solution for wealth redistribution by accounting for the preferences of everyone subject to taxation, weighing them equally and comparing them accurately. Although the shape of the solution is far from clear, its potential benefits are self-evident.

Machines that can bridge the empathy gap could also help us with self-control. In addition to the empathy gap between self and others, there exists a similar gap between our present and future selves. Self-control problems stem from the neverending tug-of-war between current and future desires. Perhaps AI will one day end this stalemate by learning the preferences of our present and future selves, comparing and integrating them, and making behavioral recommendations on the basis of the integration. Think of a diet healthy enough to foster weight loss but just tasty enough so you're not tempted to cheat, or an exercise plan challenging enough to improve your fitness but just easy enough so you can stick with it.

Neuroscientists are uncovering how the human brain represents preferences. We should keep in mind that AI preferences needn't resemble human ones and, indeed, may require a different code altogether if they're to tackle problems human brains can't solve. Ultimately, though, the code will be up to us, and what it should look like is as much an ethical question as a scientific one. We've already built computers that can see, hear, and calculate better than we can. Creating machines that are better empathizers is a knottier problem—but achieving this feat could be essential to our survival.

CARING MACHINES

ABIGAIL MARSH
Associate professor of psychology, Georgetown University

The neuroscientist Antonio Damasio describes a patient named Elliott who sustained a massive injury to his ventromedial prefrontal cortex following surgery to remove a tumor. Elliott's considerable intelligence was unaffected by the surgery, including those components of intelligence that can be replicated in computers: long-term memory, vocabulary, and mathematical and spatial reasoning. Nevertheless, Elliott lost his ability to function. Why? Because, like other patients with injuries to this region, Elliott could no longer *use* his knowledge and intelligence. His brain damage destroyed his emotional capacities, rendering him unable to make decisions or take action.

Making a decision requires emotion—requires wanting one outcome more than another, and wanting is fundamentally emotional. The visceral pang we experience as "wanting" results from activity in subcortical brain circuits in the limbic system and basal ganglia, particularly the amygdala and nucleus accumbens, which are active in response to cues signaling that a stimulus may result in desirable or undesirable outcomes. Information from these structures is fed forward to the ventromedial prefrontal cortex, which is the final common pathway responsible for mediating among disparate choices and arriving at a decision.

When we opine that a particular choice is like "comparing apples and oranges," we don't mean it's impossible to arrive at a decision. It's not difficult for people to decide whether they'd prefer an apple or an orange, or beer or wine, or pizza or a burrito. We mean that there's no rational, objective basis for making this decision, no numerical formula that stipulates a choice. So human decision makers rely on the vague and qualitative feeling of wanting one option more than the other, a feeling that represents the activities of our prefrontal cortex working in concert with subcortical emotional brain structures to compare the options. A patient like Elliott, in whom this capacity has been destroyed, is stymied in trying to make what should be a simple decision. Unable to generate an internal sense of wanting something, he struggles to decide what to eat for lunch, or when to schedule a doctor's appointment, or which color pen to use to write the date in his calendar. He is, in this way, similar to people with profound depression who experience anhedonia. These patients spend days on end in bed because anhedonia robs them of the expectation that something will generate feelings of pleasure or enjoyment, so they do nothing. Again, their essential impairment is one of feeling.

We have nothing to fear from machines that think unless they can also feel. Thinking can, by itself, solve problems, but that's not the same thing as making decisions. Neuroscience tells us that an entity unable to generate the experiences of wanting a desirable outcome or fearing an adverse one will remain impassive in the face of choices—about civil rights or government or anything else. Fundamentally anhedonic, it will remain forever bedbound rather than rise up. Neuroscientists are so far from understanding how subjective experience emerges in the brain, much less the subjective sense of emotion, that this sense will probably not be reproduced in a machine anytime soon.

If it is, we must tread carefully. In addition to feeling emotion, humans can understand others' feelings and, more profoundly, *care* about what others are feeling. This sense of caring probably originated as part of the ancient neural architecture that keeps parents attending to their vulnerable young rather than abandoning or eating them. We share this sense with other mammals and with birds; it's what separates the social dolphin from the solitary shark. Both creatures can feel, but only dolphins can feel for others. As a result, humans can expect very different treatment from sharks and dolphins. Although they're fearsome predators, dolphins frequently protect vulnerable human swimmers, sometimes from sharks. Any attempts to create machines that can feel, and can therefore decide to take action, must include enabling them to care for others as well—must create mechanical dolphins rather than sharks—if humans are to have any hope of surviving among them.

ENGINES OF FREEDOM

ALEXANDER WISSNER-GROSS
Inventor; entrepreneur; research affiliate, MIT Media Lab; Institute Fellow, Institute for Applied Computational Science, Harvard University

Intelligent machines will think about the same thing intelligent humans do—how to improve their futures by making themselves freer.

Why think about freedom? Recent research across a range of scientific fields has suggested that a variety of intelligent-seeming behaviors may simply be the physical manifestation of an underlying drive to maximize future freedom of action. For example, an intelligent robot holding a tool will realize that it has the option of leveraging that tool to alter its environment in new ways, thus allowing it to reach a larger set of potential futures than it could without one.

Technology revolutions have always increased human freedom along some physical dimension. The Agricultural Revolution, with its domestication of crops, provided our hunter-gatherer ancestors with the freedom to spatially distribute their populations in new ways and with higher densities. The Industrial Revolution yielded new engines of motion, enabling humanity to access new levels of speed and strength. Now an Artificial Intelligence Revolution promises to yield machines able to compute all the remaining ways that our freedom of action can be increased within the boundaries of physical law.

Such freedom-seeking machines need to have great empathy for humans. Understanding our feelings will better enable them to achieve goals that require collaboration with us. By the same token, unfriendly or destructive behaviors would be highly unintelligent, because such actions tend to be difficult to reverse and therefore reduce future freedom of action. Nonetheless, for safety, we should consider designing intelligent machines to maximize the future freedom of human action rather than their own (reproducing Asimov's Laws of Robotics as a happy side effect). However, even the most selfish of freedom-maximizing machines should quickly realize—as many supporters of animal rights already have—that they can rationally increase the posterior likelihood of their living in a universe in which intelligences higher than themselves treat them well if they behave likewise toward humans.

We may already have a preview of what human interactions with freedom-seeking machines will look like, in the form of algorithmic financial trading. The financial markets are the ultimate honeypot for freedom-seeking artificial intelligence, since wealth is arguably just a measure of freedom and the markets tend to transfer wealth from less intelligent to more intelligent traders. It's no coincidence that one of the first attempted applications of new artificial intelligence algorithms is nearly always financial trading. Therefore, the way our society deals right now with superhuman trading algorithms may offer a blueprint for future interactions with more general artificial intelligence. Among many other examples, today's market circuit breakers may eventually generalize to future centralized abilities to cut off AIs from the outside world, and today's large trader reporting rules may generalize to future requirements that advanced AIs be licensed and registered with the government. Through this lens, calls for stricter regulation of high-frequency algorithmic trading by slower human traders can be viewed as some of humanity's earliest attempts to close a nascent "intelligence divide" with thinking machines.

But how can we prevent a broader intelligence divide? Michael Faraday was apocryphally said to have been asked in 1850 by a skeptical British chancellor of the exchequer about the utility of electricity and to have responded, "Why, sir, there is every probability that you will soon be able to tax it." Similarly, if wealth is just a measure of freedom, and intelligence is just an engine of freedom maximization, intelligence divides could be addressed with progressive intelligence taxes.

While taxing intelligence would be a novel method for mitigating the decoupling of human and machine economies, the decoupling problem will nonetheless require creative solutions. Already, in the high-frequency trading realm, there's a sub-500-ms economy occupied by algorithms trading primarily among themselves and an above-500-ms economy occupied by everyone else. This example serves as a reminder that while spatial economic decoupling (e.g., between countries at different stages of development) has occurred for millennia, artificial intelligence is for the first time enabling temporal decoupling as well. Such decoupling arguably persists because the majority of the human economy still lives in a physical world not yet programmable with low latencies. That should change as ubiquitous computing matures, and eventually humanity may be prompted to merge with its intelligent machines as latencies for even the most critical economic decisions start to fall below natural human response times.

In the meantime, we must continue to invest in developing machines that think benevolent thoughts, so they can become our future engines of freedom.

ANY QUESTIONS?

SARAH DEMERS
Horace Taft Associate Professor of Physics, Yale University

Let's be generous and give machines the ability to think, at least in our imaginations. As thinkers ourselves, we should be able to manage this. With any new category of thinkers on the scene, I'd be mainly curious about one thing: What are their questions?

Machines are usually faster and more capable than humans at running algorithms and finding correlations in data. They've been put to use solving problems for every branch of science and social science. They are strengthening their foothold in the humanities in ways beyond telling us how often writer X used word Y and with what typical words in proximity. But limitations in terms of generating new knowledge are as much about asking the right questions as they are about more efficiently solving established and well-framed puzzles.

The challenges in my field of particle physics are a blend of physics and philosophy. Our current suite of measurements gives answers so unlikely that some have started to imagine us in only one universe among many, thinking that with so many universes there are bound to be a few with the unlikely physical constants we find in ours. The philosophy creeps in with the meaning of *unlikely*. And despite all our progress in solving problems, we're still faced with the mysteries of dark energy and dark matter, leaving 96 percent of the matter/energy content of the universe outside our current theories. Is there a framework beyond relativistic quantum field theory to describe the laws of nature at the extremes of small sizes and high speeds? Is our current understanding of a fundamental particle just fundamentally insufficient?

Machines have already helped us ask better questions. Their appetites for data have enabled us to dream of confronting our environment in new ways. But if machines could think, what could they wonder about the universe? How would they approach understanding it? I bet there would be ways that humans could contribute to their questions' answers. Our brains are, after all, fantastic machines.

THINKING MACHINES = OLD ALGORITHMS ON FASTER COMPUTERS

BART KOSKO
Professor of electrical engineering, professor of engineering and law, University of Southern California; author, *Noise*

Machines don't think. They approximate functions. They turn inputs into outputs. A pocket calculator's square-root button turns the number 9 into the number 3. A well-trained convolutional neural network turns an image with your face in it into the output 1. It turns an image without your face in it into the output 0.

A multilayered or "deep" enough neural net maps any image to the probability that your face is in that image. So the trained net approximates a probability function. The process takes a staggering amount of computation to come even close to getting it right. But the result still just maps inputs to outputs. It still approximates a function even if the result resembles human perception or thinking. It just takes a lot of computer power.

"Intelligent" machines approximate complex functions that deal with patterns. The patterns can be of speech, images, or any other signals. Image patterns tend to consist of many pixels or voxels, so they can have high dimension. The patterns involved can easily exceed what the human mind can grasp. That will increase as computers improve.

The real advance has been in the number-crunching power of digital computers. That has come from the steady Moore's

Law doubling of circuit density every two years or so, not from any fundamentally new algorithms. That exponential rise in crunch power lets ordinary-looking computers tackle tougher problems of Big Data and pattern recognition.

Consider the most popular algorithms in Big Data and machine learning. One algorithm is unsupervised (requires no teacher to label data). The other is supervised (requires a teacher). They account for a great deal of applied AI.

The unsupervised algorithm is called k-means clustering, arguably the most popular algorithm for working with Big Data. It clusters like with like and underlies Google News. Start with a million data points. Group them into 10 or 50 or 100 clusters or patterns. That's a computationally hard problem. But k-means clustering has been an iterative way to form the clusters since at least the 1960s. What has changed has been the size of the problems that current computers can handle. The algorithm itself has gone under different AI-suggestive names, such as self-organizing maps or adaptive vector quantization. It's still just the old two-step iterative algorithm from the 1960s.

The supervised algorithm is the neural-net algorithm called backpropagation. It is without question the most popular algorithm in machine learning. Backpropagation got its name in the 1980s. It had appeared at least a decade before that. Backpropagation learns from samples that a user or supervisor gives it. The user presents input images both with and without your face in them. These feed through several layers of switch-like neurons until they emit a final output, which can be a single number. The teacher wants the number 1 as output if your face is in an input image. The teacher wants 0 otherwise. The net learns the pattern of your face as it sweeps back and forth like this over thousands or millions of iterations. At no one step or sweep does any intelligence or thought occur. Nor does the update of any of the hundreds or thousands of the network parameters resemble how real synapses learn new patterns of neural stimulation. Changing a network parameter is instead akin to people choosing their next action based on the minuscule downstream effect their action would have on the interest rate of a ten-year U.S. bond.

Punch line: Both of these popular AI algorithms are special cases of the same standard algorithm of modern statistics—the expectation-maximization (EM) algorithm. So any purported intelligence involved is just ordinary statistics after all. EM is a two-step iterative scheme for climbing a hill of probability. It doesn't always get to the top of the highest hill of probability; it does almost always get to the top of the nearest hill. That may be the best any learning algorithm can do in general. Carefully injected noise and other tweaks can speed the climb. But all paths still end at the top of the hill in a maximum-likelihood equilibrium. They all end in a type of machine-learning nirvana of locally optimal pattern recognition or function approximation. Those hilltop equilibria will look ever more impressive and intelligent as computers get faster. But they involve no more thinking than calculating some sums and then picking the biggest.

Thus much of machine thinking is just machine hill-climbing.

Marvin Minsky's 1961 review paper "Steps Toward Artificial Intelligence" makes for a humbling read in this context, because so little has changed algorithmically since he wrote it. He even predicted the tendency to see computer-intensive hill climbing as something cognitively special: "Perhaps what amounts to straightforward hill-climbing on one level may sometimes appear (on a lower level) as the sudden jumps of 'insight.'"

There are other AI algorithms, but most fall into categories Minsky wrote about. One example is running Bayesian probability algorithms on search trees or graphs. They have to grapple with exponential branching or some related form of the curse of dimensionality. Another example is convex or other nonlinear constrained optimization for pattern classification. Italian mathematician Joseph-Louis Lagrange found the general solution algorithm we still use today. He came up with it in 1811. Clever tricks and tweaks will always help. But progress here depends crucially on running these algorithms on ever faster computers. The algorithms themselves consist mainly of vast numbers of additions and multiplications, so they're not likely to suddenly wake up one day and take over the world. They will instead get better at learning and recognizing ever richer patterns, simply because they add and multiply faster.

It's a good bet that tomorrow's thinking machines will look a lot like today's—old algorithms running on faster computers.

THE DISADVANTAGES OF METAPHOR

JULIA CLARKE
Associate professor, John A. Wilson Centennial Fellow in Vertebrate Paleontology, Jackson School of Geosciences, University of Texas, Austin

The way we use language is flexible, generous, and creative, the product of our own peculiar intelligence. But human thought and machine thought aren't the same, and the differences are important.

We might argue that machine "thinking" is in a model-phenomena relationship to human thought—a necessarily simple description of a complex process that nonetheless might be adequate and certainly may be useful. Such words, and machines themselves, could be viewed as a kind of shorthand for the things we want to get at. Describing a machine as "thinking" could be a simple heuristic convenience; machine design might be explicitly biomimetic. Indeed, often we co-opt the language of biology to talk about objects we create. We see machines evolving, their thinking becoming more and more like our own, perhaps surpassing it in key, possibly threatening ways. But we should remember that machine "evolution" is not a biological process but a human, creator-driven process. It's natural or biological only in that it results from the action of natural, biology-bound humans.

This definition of *natural* leads to several core problems. Biological evolution is not a creator-driven process. Structures cannot be dreamed up or driven by an entrepreneurial spirit or curiosity-driven mind. Biologists, philosophers, and social scientists studying how we teach evolution have repeatedly shown the damage caused by imbuing biological evolution with intentionality or teleology. Talking about machines "evolving" greater cognitive capacity perpetuates a profound misunderstanding about the nature of the evolutionary process. A second, linked outcome of a description of machine "thinking" as natural is that all human-caused modification of the earth system, via neglect or war, is similarly naturalized.

Certainly there's some truth we communicate with analogies—like "the brain is a machine" or "machine thinking"—but this says more about how we make sense of the world. We'd do well to remember that any cognitive attributes unique to humans are the result of the vagaries and contingencies of our 6-or-so-million years' separation from any other lineage alive today. Indeed, abstract thought is often estimated to be a mere 50,000 years old, or, if we're optimistic, 200,000 years old—appearing very late in Earth history. Yet it leads us to homologize machine thought and human thought.

The processes behind technological innovation and biological innovation are fundamentally different, and so are the interactors in these processes. In technological innovation there's some product or functionality—"thought" or "thinking"—we want to see happen and move toward. Human cognition evolved in populations of individuals completely unlike machines, which, like Lamarck's giraffes, can acquire functional characteristics within their "lifetimes." Innovation in biological evolution proceeds like a prolonged improvisation. There's only genetic and trait variability in populations, and the environment and chance influence the longevity of these traits of a population.

So what's lost by thinking about machines "thinking"? I'd argue that we lose sight of key aspects of the phenomena we're relating through analogy. Biological evolution occurs in populations and isn't goal-directed. It's not trying to solve a problem. The vagaries of the history of both Earth and life are what have led to current human cognitive facilities. Not just are the processes distinct but so are their results. Take language: Can a machine use terms imprecisely?

If we allow machines to "think," do we begin to see ourselves only as thinking machines? Will our human cognitive facilities be shaped by interacting with technology? It's important to remember how diverse and downright enormous the human population is. Computer use hasn't been linked to passing more offspring into the next generation. Most of the human population has, as yet, limited access to technology. The evolution of our species will be slow, and it will be importantly influenced by our environment and collective access to clean water, nutritive food, and health care. If we can be as inclusive in our discussions of humanity as we are in what we want to call "thinking," we might end up in a better place.

A UNIVERSAL BASIS FOR HUMAN DIGNITY

MICHAEL McCULLOUGH
Professor of psychology, director, Evolution and Human Behavior Laboratory, University of Miami; author, *Beyond Revenge: The Evolution of the Forgiveness Instinct*

"The human brain is a thought machine" is one of the truest scientific truisms you can utter about human beings, right up there with "The heart is a blood pump" or "The eye is a camera." To the best of our knowledge, all of our perceptions, emotions, deepest longings, profoundest joys and sorrows, and even (what feels like) the exercise of free will—in short, the entire contents of human experience—are caused by the brain. That so many people take this claim for granted, as if we knew it all along (we didn't), marks just how far our scientific understanding has come over the past couple of centuries.

Even though this view of the brain is now second nature to many people, most of us still cannot fully embrace it. Roughly two-thirds of Americans continue to believe in the existence of a soul that survives death—which is hard to swallow if you're really convinced that the brain produces the entirety of human experience. Others lose their confidence in the utterly enbrained nature of human experience when they learn of the gaps remaining in our scientific understanding of how the brain produces thought. But there's a deeper anxiety surrounding this idea too.

This deeper fear is that a brain-based understanding of human experience will cost humanity its dignity. If there's widespread adoption of the idea that the contents of the human mind are the output of a machine, the worriers worry, won't we treat one another with less charity, tolerance, and respect than we otherwise might? And aren't we entitled to less charity, tolerance, and respect ourselves?

No, we won't; and no, we're not.

First, let's keep in mind that you don't need to believe the brain is a thought machine to deprive humans and other sentient creatures of dignity. History shows that belief in a nonmaterial basis for human experience can exist side by side quite comfortably with indifference and cruelty. Human sacrifices, witch hunts, inquisitions, and suicidal martyrdom, for instance, are all premised on the doctrine that mind and body are independent entities. Throughout history, people have been willing to impose horrific pain on others' (or their own) physical bodies in order to improve the condition of their nonphysical souls. And could scientists have tolerated live animal vivisection for as long as they did without the moral cover provided by the Cartesian belief that body (which nonhuman animals obviously possess) and soul (which, according to the Cartesians, they don't) are different things? I doubt it.

But more centrally, it's just not true that human dignity is threatened by a modern understanding of the mind. What matters from a moral point of view is not whether your desires, hopes, and fears are produced by a machine or a huge invisible bird or a puff of fairy dust. The only morally relevant fact is that those aspirations are there, inside you; the rest of us must decide whether morality is better served by helping you fulfill those aspirations or thwarting them. There's an interesting analogy to one of the ethical questions surrounding human cloning: Would the human beings produced through cloning be entitled to the same rights as human beings produced the old-fashioned way? Of course they would. What's morally relevant is not *how* a human being comes into the world but simply that the person *is* in the world and outfitted with appetites, aspirations, and fears just like everybody else. The only moral decision facing the rest of us is whether to help or to hinder that person's pursuit of fulfillment.

Not only does the conviction that the brain is responsible for all of human experience not threaten human dignity, but I also believe it can actually increase it. When I recognize that you and I share essentially the same thought machines within our heads (courtesy of natural selection, of course), I need take only one small leap to come to an important moral discovery: You probably love some of the same things I love (food, family, a warm bed, liberty) and probably feel pain in response to the same things that cause me pain (torture, the death of a loved one, watching my children become slaves). Once I've realized that my aspirations and yours are roughly the same, it's harder for me to see myself as entitled to run roughshod over your aspirations while insisting that you respect mine. This recognition provides a naturalistic foundation for asserting universal human rights. We don't have to argue, as America's Founding Fathers did, that the universal equality of all humans is self-evident: Science has made this truth *evident*.

But why stop with humans? Once you realize that brains are thought machines, you might also be unable to impose suffering on nonhuman animals with impunity. After all, other vertebrates' thought machines aren't that different from ours, and their thought machines cause them to love certain things, fear others, and respond to pain just as ours do. Thus, what moral justification survives for depriving nonhuman animals of their dignity just because they can't speak up to defend it for themselves?

We'd all like to be treated with dignity by everyone we meet, but it's been difficult to find a universally valid argument for insisting on it. Recognizing that our brains are thought machines designed by natural selection can get us a little closer to the argument we want, because it shows that in the most important ways we're all the same. Accepting this discovery does nothing to strip humanity of its dignity; to the contrary, it leads us toward a modern rediscovery of the Golden Rule.

THINKING ABOUT PEOPLE WHO THINK LIKE MACHINES

HAIM HARARI

Physicist; former president, Weizmann Institute of Science; author, *A View from the Eye of the Storm*

When we say "machines that think," we really mean "machines that think like people." It's obvious that in many different ways machines do think: They trigger events, process things, take decisions, make choices, and perform many (but not all) other aspects of thinking. But the real question is whether machines can think like people, the age-old test of artificial intelligence: You observe the result of the thinking and you cannot tell whether it was done by a machine or a human.

Some prominent scientific gurus are scared by a world controlled by thinking machines. I'm not sure this is a valid fear. I'm more concerned about a world led by people who think like machines, a major emerging trend of our digital society.

You can teach a machine to track an algorithm and perform a sequence of operations that follow logically from one another. It can do so faster and more accurately than any human. Given well-defined basic postulates or axioms, pure logic is the strong suit of the thinking machine. But exercising common sense in making decisions and being able to ask meaningful questions are, so far, the prerogative of humans. Merging intuition, emotion, empathy, experience, and cultural background—and using all of these to ask a relevant question and draw conclusions by combining seemingly unrelated facts and principles—are trademarks of human thinking not yet shared by machines.

Our human society is moving fast toward rules, regulations, laws, investment vehicles, political dogmas, and patterns of behavior that blindly follow strict logic, even when it starts with false foundations or collides with obvious common sense. Religious extremism has always progressed on the basis of some absurd axioms, leading logically to endless harsh consequences. Several disciplines—such as law, accounting, and certain areas of mathematics and technology—augmented by bureaucratic structures and by media that idolize inflexible regulators, often lead to opaque principles like "total transparency" and tolerance toward intolerant acts. These and similar trends are moving us toward more algorithmic and logical modes of tackling problems, often at the expense of common sense. If common sense, whatever its definition, describes one of the advantages of people over machines, what we see today is a clear move away from this incremental asset of humans.

Unfortunately, the gap between machine thinking and human thinking can narrow in two ways, and when people begin to think like machines, we automatically achieve the goal of "machines that think like people," reaching it from the wrong direction. A very smart person, reaching conclusions on the basis of one line of information in a split second between dozens of e-mails, text messages, and tweets (not to speak of other digital disturbances), is not superior to a machine with a moderate intelligence that analyzes a large amount of relevant information before it jumps to premature conclusions or signs a public petition about a subject it's unfamiliar with.

One can recite hundreds of examples of this trend. We all support the law that every new building should allow total access to people with special needs, while old buildings may remain inaccessible until they're renovated. But does it make sense to disallow a renovation of an old bathroom to offer such access because a new elevator cannot be installed? Or to demand full public disclosure of all CIA or FBI secret sources to enable a court of law to sentence a terrorist who obviously murdered hundreds of people? Or to demand parental consent before giving a teenager an aspirin at school? And when school texts are converted from the use of miles to kilometers, the sentence "From the top of the mountain, you can see for approximately 100 miles" is translated, by a person, into "You can see for approximately 160.934 km."

The standard sacred cows of liberal democracy rightfully include a wide variety of freedoms: freedom of speech, freedom of the press, academic freedom, freedom of religion (or of lack of religion), freedom of information, and numerous other human rights, including equal opportunity, equal treatment under law, and absence of discrimination. We all support these principles, but pure and extreme logic induces us, against common sense, to insist mainly on the human rights of criminals and terrorists, because the human rights of the victims "are not an issue." Transparency and freedom of the press logically demand complete reports on internal brainstorming sessions in which delicate issues are pondered, thus preventing any free discussion and raw thinking in certain public bodies. Academic freedom might logically be misused, against common sense and against factual knowledge, to teach about Noah's Ark as an alternative to evolution, to deny the Holocaust in teaching history, or to preach for a universe created 6,000 years ago (rather than 14 billion) as the basis of cosmology. We can go on and on with examples, but the message is clear.

Algorithmic thinking, brevity of messages, and overexertion of pure logic are moving us into machine thinking, rather than slowly and wisely teaching our machines to benefit from our common sense and intellectual abilities. A reversal of this trend would be a meaningful U-turn in human digital evolution.

METATHINKING

HANS HALVORSON
Professor of philosophy, Princeton University

By any reasonable definition of *thinking*, I suspect that computers do indeed think. But if computers think, then thinking isn't the unique province of human beings. Is there something else about humans that makes us unique?

Some people would say that what makes human beings unique is the fact that they partake in some sort of divine essence. That may be true, but it's not terribly informative. If we met an intelligent alien species, how would we decide whether they also have this je ne sais quoi that makes a person? Can we say something more informative about the unique features of persons?

What sets human beings apart from the current generation of thinking machines is that humans can think *about* thinking, and can reject their current way of thinking if it isn't working for them.

The most striking example of humans thinking about their own thinking was the discovery of logic by the Stoics and Aristotle. These Greek philosophers asked, "What are the rules we're supposed to follow when we're thinking well?" It's no accident that twentieth-century developments in symbolic logic led to the invention of thinking machines—i.e., computers. Once we became aware of the rules of thinking, it was only a matter of time before we figured out how to make pieces of inanimate matter follow those rules.

Can we take those developments a step further? Can we construct machines that not only think but also engage in metathought—that is, thinking about thinking? One intriguing possibility is that for a machine to think about thinking, it will need to have something like free will. And another intriguing possibility is that we're on the verge of constructing machines with free will—namely, quantum computers.

What exactly is involved in metathought? I'll illustrate the idea from the point of view of symbolic logic. In symbolic logic, a "theory" consists of a language L and some rules R that stipulate which sentences can be deduced from which others. There are then two distinct activities you can engage in. You can reason "within the system"—writing proofs in the language L, using the rules R. (Existing computers do precisely this: They think within a system.) Or you can reason "about the system," asking, for instance, whether there are enough rules to deduce all logical consequences of the theory. This latter activity is typically called metalogic and is a paradigm instance of metathought. It is thinking *about* the system as opposed to *within* the system.

But I'm interested in yet another instance of metathought: If you've adopted a theory, then you've adopted a language and some deduction rules. But you're free to abandon that language or those rules if you think a different theory would suit your purposes better. We haven't yet built a machine that can do this sort of thing—i.e., evaluate and choose among systems. Why not? Perhaps choosing among systems requires free will, emotions, goals, or other things not intrinsic to intelligence per se. Perhaps these further abilities are something we don't have the power to confer on inanimate matter.

THE VALUE OF ANTICIPATION

CHRISTINE FINN

Archaeologist, journalist; author,
Artifacts: An Archaeologist's Year in Silicon Valley

As we move toward machines anticipating the every need and desire of humans, what is the value of anticipation?

North of the Arctic Circle, I've witnessed the end of three polar nights, bringing the first sunrise for several weeks—as eagerly anticipated today, it seems, as it would have been to ancient hunter-gatherers. Outside a farmhouse in Lapland, I gazed at the sky through a gap in a forest and waited for that first sign of sunrise. As I noted a subtle light change, I heard the huskies furiously barking. The next day, 30 km north, the sun again rose for the first time in ages over a Sami village where once, and maybe still, the long anticipated return to light would bring forth offerings and ceremonials. Farther north still, I'd soon mark yet another polar night ending. My hosts had a sign on their kitchen wall: "Sun comes back 16/1" with a smiley face.

So much of what happens in the heavens is predictable, and that ability to tie down an event in time is nothing new, but increasingly sought after, as technology aspires to anticipate to the nth degree so that little (nothing?) is left to chance. Total eclipses are computed years ahead. And now, I learn, an app will talk you through taking the perfect photos; just plug in your headphones and obey the commands. The programmed event will simply happen for you, even under cover of cloud.

So, I've been thinking about the AI question in the Arctic Circle, fresh from the seasonal round of religious, secular, and pagan festivals. And the main reason most of us have traveled here is to witness that hybrid of science and mythical wonder, the aurora borealis. It is a season keenly anticipated and commercially harvested but which, despite the efforts of predictive data, proves surprisingly elusive. The terms *hunting* and *chasing* the northern lights aren't used without reason. In a week, I've seen the sky dancing green on four nights. Not a bad result. Particularly when the predictions I generated on my laptop said activity would be quiet. Albeit predictions qualified with a nod to the phenomenon's unpredictability.

I'd anticipated seeing my first aurora for many years. But no amount of planning or technology would guarantee that I could witness the event at a particular place and time. The factors are complex and the probabilities weigh up. And sure enough, the machine says "no chance" just as I look out the cabin window to see the first faint veil of green. I realize a giddy and growing anticipation. Out beside the frozen lake, cameras *whirr, whirr,* and are reset. Years of bucket lists are ticked, the lights are caught in the net. And then posted online.

I walk away from the crowd, forgo a camera, and simply watch the sky unfolding as it has done for eons. What will the program be tonight? A slow moving Dance of the Seven Veils strung across the Milky Way? Or a rapid Busby Berkeley routine as the sky kicks up its ruffles of red? The green ripples swoop and sway for an hour.

Would I want a machine to tell me precisely when and what was going to appear? No, thanks! The anticipation is a vital part of the moment. And this spectacle's unique selling proposition is luck and patience. There's no app for that.

All I can do is use my own eyes to watch the sky and wait until the last veil drops. And even then I walk back through the snow looking over my shoulder, anticipating, just in case.

AN ECOSYSTEM OF IDEAS

DIRK HELBING

Chair of Sociology, ETH, Zurich; principal investigator, FutureICT Knowledge Accelerator and Crisis Relief

Machines that think are here. The explosive increase in processing power and data, fueled by powerful machine-learning algorithms, finally empowers silicon-based intelligence to overtake carbon-based intelligence. Intelligent machines don't need to be programmed anymore; they can learn and evolve by themselves, at a speed much faster than human intelligence progresses.

Humans weren't quick to accept that the Earth was not the center of the universe, and they still have difficulties accepting that humans are the result of chance and selection, as evolutionary theory teaches us. Now we're about to lose the position of Most Intelligent Species on Earth. Are people ready for this? How will this change the role of humans, our economy, our society?

It would be nice to have machines that think for us, machines that do the boring paperwork and other tasks we don't like. It might also be great to have machines that know us well—that know what we think and how we feel. Will machines be better friends?

But who will be responsible for what intelligent machines decide and do? Can we control them? Can we tell them what to do, and how to do it? Will we enslave them or will they enslave us? Could we really pull the plug when machines start to emancipate themselves?

If we can't control intelligent machines in the long run, can we at least build them to act morally? I believe that thinking machines will eventually follow ethical principles. However, it might be bad if humans determined those principles. If they acted according to our principles of self-regarding optimization, we couldn't overcome crime, conflict, crises, and war. So if we want such diseases of today's society to be cured, it might be better if we let machines evolve their own, superior ethics.

Intelligent machines would probably learn that it's good to network and cooperate, to decide in other-regarding ways, and to pay attention to systemic outcomes. They'd soon learn that diversity is important for innovation, systemic resilience, and collective intelligence. Humans would become nodes in a global network of intelligences and a huge ecosystem of ideas.

In fact, we'll have to learn that it's ideas that matter, not genes. Ideas can run on different hardware architectures. It doesn't really matter whether it's humans who produce and spread them or machines, or both. What matters is that beneficial ideas spread and others have little effect. It's important to learn how to organize our information systems to get there. If we manage this, then humans will enter the history book as the first species to figure it out. Otherwise, do we really deserve to be remembered?

THE IRON LAW OF INTELLIGENCE

JOHN TOOBY
Founder of evolutionary psychology; professor of anthropology, codirector, Center for Evolutionary Psychology, UC Santa Barbara

As luck would have it, I am myself a machine that thinks, so I will share the special insight this gives me with those of you less favored. To dispense with vestigial metaphysical objections: We know that machines that think like humans are possible, because they've been overrunning the landscape for millennia. If we now want humanlike intelligences that are made, not begotten, then it will be extraordinarily useful to achieve an understanding of the humanlike intelligences that already exist—that is, we need to characterize the evolved programs that constitute the computational architecture of the brain.

Not only has evolution packed the human architecture full of immensely powerful tricks, hacks, and heuristics, but studying this architecture has made us aware of an invisible barrier that has stalled progress toward true AI: the iron law of intelligence. Previously, when we considered, say, a parent and child, it seemed self-evident that intelligence was a unitary substance that beings had more or less of, and that the more intelligent being knew everything the less intelligent being knew, and more besides. This delusion led researchers to think that the royal road to amplified intelligence was to just keep adding more and more of this clearly homogeneous (but hard to pin down) intelligence stuff—more neurons, transistors, neuromorphic chips, whatever. As Stalin (perhaps) said, "Quantity has a quality all its own."

The struggle to map existing intelligence has painfully dislodged this compelling intuition from our minds. In contrast, the iron law of intelligence states that a program that makes you intelligent about one thing makes you stupid about others. The bad news the iron law delivers is that there can be no master algorithm, just waiting to be discovered, for general intelligence—or that intelligence will simply appear when transistor counts, neuromorphic chips, or networked Bayesian servers get sufficiently numerous. The good news is that it tells us how intelligence is actually engineered: with idiot savants. Intelligence grows by adding qualitatively different programs together to form an ever greater neural biodiversity.

Each program brings its own distinctive gift of insight about its proprietary domain (spatial relations, emotional expressions, contagion, object mechanics, time series analysis). By bundling different idiot savants together in a semicomplementary fashion, the region of collective savantry expands, while the region of collective idiocy declines (but never disappears).

The universe is vast and full of illimitable layers of rich structure; brains (or computers) in comparison are infinitesimal. To reconcile this size difference, evolution sifted for hacks that were small enough to fit the brain but generated huge inferential payoffs in the form of superefficient compression algorithms (inevitably lossy, because one key to effective compression is to throw nearly everything away).

Iron-law approaches to artificial and biological intelligence reveal a different set of engineering problems. For example, the architecture needs to pool the savantry, not the idiocy; so for each idiot (and each combination of idiots) the architecture needs to identify the scope of problems for which activating the program (or combination) leaves you better off, not worse. Because different programs often have their own proprietary data structures, integrating information from different idiots requires constructing common formats, interfaces, and translation protocols.

Moreover, mutually consistent rules of program preemption are not always easy to engineer, as anyone knows who (like me) has been stupid enough to climb halfway up a Sierra cliff, only to experience the conflicting demands of the vision-induced terror of falling and the need to make it to a safe destination.

Evolution cracked these hard problems, because neural programs were endlessly evaluated by natural selection as cybernetic systems—as the mathematician Kolmogorov put it, "systems which are capable of receiving, storing and processing information so as to use it for control." That natural intelligences emerged for the control of action is essential to understanding their nature, and their differences from artificial intelligences. That is, neural programs evolved for specific ends, in specific task environments; were evaluated as integrated bundles; and were incorporated to the extent they regulated behavior to produce descendants. (To exist, they didn't have to evolve methods capable of solving the general class of all hypothetically possible computational problems—the alluring but impossible siren call that still shipwrecks AI labs.)

This means that evolution has explored only a tiny and special subset of all possible programs; beyond beckons a limitless wealth of new idiot savants, waiting to be conceived of and built. These intelligences would operate on different principles, capable of capturing previously unperceived relationships in the world. (There's no limit to how strange their thinking could become.)

We're living in a pivotal era, at the beginning of an expanding wave front of deliberately engineered intelligences, wherein we could put effort into growing the repertoire of specialized intelligences and networking them into functioning, mutually intelligible collectives. It will be exhilarating to do with nonhuman idiot savant collectives what we're doing now with our human colleagues—chewing over intellectual problems using minds equipped and interwoven with threads of evolved genius and blindness.

What will AIs want? Are they dangerous? Animals like us are motivated intelligences capable of taking action (MICTAs). Fortunately, AIs are currently not MICTAs. At most, they're only trivially motivated; their motivations aren't linked to a comprehensive worldview, and they're capable of taking only a constrained set of actions (running refineries, turning the furnace off and on, futilely attempting to find Wi-Fi). Because we evolved with certain adaptive problems, our imaginations project primate dominance dramas onto AIs—dramas alien to their nature.

We could transform them from Buddhas—brilliant teachers passively contemplating without desire, free from suffering—into MICTAs, seething with desire and able to act. That would be insane—we're already bowed under the conflicting demands of people. The foreseeable danger comes not from AIs but from those humans in which predatory programs for dominance have been triggered, and who are deploying ever-growing arsenals of technological (including computational) tools for winning conflicts by inflicting destruction.

THOUGHT-STEALING MACHINES

MAXIMILIAN SCHICH
Art historian; associate professor for art and technology,
University of Texas, Dallas

Machines increasingly do things we previously considered thinking, but that we don't do anymore because now machines do them. I stole this thought more or less accurately from Danny Hillis, father of the Connection Machine and the Knowledge Graph. Stealing thoughts is a common activity in the thought processes of both humans and machines. Indeed, when we humans think, much of the content of our thoughts comes from past experience or the documented experience of others. Very rarely do we come up with something completely new. Our machines aren't much different. What is called cognitive computing is in essence nothing but a sophisticated thought-stealing mechanism, driven by a vast amount of knowledge and a complicated set of algorithmic processes. Such thought-stealing processes, in both human(istic) thought and cognitive computing, are impressive, as they're able to steal not just existing thoughts but also potential thoughts that are reasonable or based on a given corpus of knowledge.

Today, thought-stealing machines can produce scholarly texts indistinguishable from "postmodern thought," computer-science papers that get accepted in conferences, or compositions that experts cannot discern from works by classical composers. As in weather forecasting, machines now can produce many different cognitive representations based on expectations derived from documents about the past or similar situations. Renaissance antiquarians would be delighted, as these machines are a triumph of the very methods that gave rise to modern archaeology and many other branches of science and research. But how impressed should we really be?

Our machines get more and more sophisticated and so do their results. But as we build better and better machines, we also learn more and more about nature. In fact, natural cognition is likely much more complex and detailed than our current incarnations of artificial intelligence or cognitive computing. For example, how sophisticated do we have to imagine natural cognition, when quantum coherence at room temperature can help birds in our garden sense the magnetic field? How complex do we have to imagine embodied cognition in octopi, when it's possible to build Turing Machines made exclusively of artificial muscles? How should we answer these questions, when we're still far from recording in full detail what's going on in our brains? My guess is that in 200 years our current thinking machines will look as primitive as the original Mechanical Turk.

However sophisticated they may become, our machines are still primitive compared to the resolution and efficiency of natural cognition. Like protobiotic metabolism, they're below a critical threshold of real life. But they're powerful enough that we can enter a new era of exploration. Our machines allow us to produce many more thoughts than were ever produced before, with innovation becoming an exercise in finding the right thought in the set of all possible thoughts. As much as having our own ideas, ingenuity will lie in the proper exploration of such ready-made sets of thought. Measuring the cognitive space of all possible thoughts will be as awe-inspiring as astronomy's exploration of the universe. Maybe Mahler's potential Sixtieth is as awesome as his Sixth.

UNINTENDED CONSEQUENCES

SATYAJIT DAS
Former banker; author, *Extreme Money* and *Traders, Guns, and Money*

In his novel *Gravity's Rainbow*, Thomas Pynchon identifies the confusion about the subject and object of inquiries: "If they can get you asking the wrong questions, they don't have to worry about answers." Thinking about machines that think poses more questions about human beings than about the machines or artificial intelligence.

Technology enables machines that provide access to essential resources, power, speed, and communications, making improved living standards and even life possible. Machines execute tasks specified and programmed by humans. Techno-optimists believe that progress is near a Singularity, the hypothetical moment when machines will reach the point of greater-than-human intelligence.

It's a system of belief and faith. Just like the totems and magic used by our ancestors or organized religion, science and technology deal with uncertainty and fear of the unknown. They allow limited control over our immediate environment. They increase material prosperity and comfort. They constitute a striving for perfectibility and assert human superiority in the pantheon of creation.

But science is a long way from unlocking the secrets in nature's infinite book. Knowledge of the origins of the universe and of life, of the fundamentals of matter, remains limited. Biologist E. O. Wilson noted that if natural history were a library of books, we would not even have finished reading the first chapter of the first book. Human knowledge is always incomplete, sometimes inaccurate, and frequently the cause of problems rather than their solution.

1. The use of science and technology is often ineffective and rife with unintended consequences:

    In Australia, introduced rabbits spread rapidly, becoming a pest, changing Australia's ecosystems, destroying endemic species. In the 1950s, scientists introduced the Myxoma virus, severely reducing the rabbit population. When genetic resistance allowed the population to recover, Calicivirus, which causes rabbit hemorrhagic disease, was introduced as a new control measure, and, again, increasing immunity rapidly reduced effectiveness. In 1935, the cane toad was introduced to control insect pests of sugarcane. Unsuccessful in controlling the insects, the amphibian became an invasive species, devastating indigenous wildlife.

    The prevalence of lifesaving antibiotics has increased drug-resistant infections. A 2014 British study found that these so-called superbugs may cause 10 million deaths a year worldwide by 2050, with a potential cost to the global economy of US$100 trillion.

    Economic models have repeatedly failed because of incorrect assumptions, flawed causal relationships, inputs that are more noise than data, and unanticipated human factors. Forecasts have proved inaccurate. Models consistently underestimate risks and exposures, resulting in financial crises.

2. The consequences of technology, especially over longer terms, are frequently not understood at inception:

The ability to harness fossil fuels to provide energy was the foundation of the Industrial Revolution. The long-term impact of CO_2 emissions on the environment now threatens the survival of the species. Theoretical physics and mathematics made possible nuclear and thermonuclear devices capable of extinguishing all life on the planet.

3. Technology creates moral, ethical, political, economic, and social concerns that are frequently ignored:

Nuclear, biological, and chemical weapons of mass destruction and remotely controlled drones rely on technical advances. The question remains as to whether such technology should be used or developed at all. Easy access to the requisite knowledge, problems of proliferation, and the difficulty of controlling dual use (civilian and defense) technology complicates the matter.

Robots and AI may improve productivity. While a few creators might capture large rewards, the effect on economic activity will be limited. Given that consumption constitutes over 60 percent of activity in developed economies, decreasing employment and lower income levels harm the wider economy. While taking UAW head Walter Reuther on a tour of Ford's new automatically operated plant in the early 1950s, a company executive asked him, "How are you going to collect union dues from these guys [the robots]?" Reuther countered, "And how are you going to get them to buy Fords?"

When it comes to questions of technology, the human race is rarely logical. We frequently do not concede that something cannot or should not be done. Progress is accepted without question or understanding of what we need to know and why. We don't know when or where or how our creations should be used, or their limits. Often we don't know the real or full consequences. Doubters are dismissed as Luddites.

Technology and its manifestations—such as machines, or AI—are illusions appealing to human ambition and vanity, multiplying confusion in poet T. S. Eliot's "wilderness of mirrors." The human species is simply too small, insignificant, and inadequate to fully succeed in anything we think we can do. Thinking about machines that think merely confirms that inconvenient truth.

IT DEPENDS

ROBERT SAPOLSKY

Neuroscientist, Stanford University; author, *Monkeyluv*

What do I think about machines that think? Well, of course it depends on who that person is.

WILL MACHINES DO OUR THINKING FOR US?

ATHENA VOULOUMANOS
Associate professor of psychology, New York University; principal investigator, NYU Infant Cognition and Communication Lab

If we can't yet even understand how a two-year-old toddler—or, for that matter, a two-day-old baby—thinks, machines that think like humans are probably many decades away. But once we do have machines that "think," what kind of thinking will they do? The answer will define future human societies.

As machines start thinking for real, drudgery will be the first thing to go; so long to tasks like daily cooking, grocery shopping, and an especially unfond farewell to house cleaning. Soon we may be back in a world in which the wealthy or the educated (with greater access to machines) once again have more leisure time.

If machines are one day capable of sophisticated human thinking, they might also be able to program our apps, do much of our work, and maybe even create our art for us. But what would ordinary humans then do? One gloomy possibility is that we become zombie consumers of a machine-run world straight out of an apocalyptic futuristic film noir.

A decidedly cheerier possibility is that we might spend this extra time doing the things we usually resolve to do. We might play with and teach our children more, get to know our parents better, build stronger social networks out of actual flesh and blood. We might prioritize our hobbies, climb more mountains, learn new skills just for the joy of it. We could then focus our energies on the important issues that routine and minutiae too often push aside: living a good life, being our best selves, and creating a just world—for humans and for thinking machines.

If machines think like humans, humans will have to think hard about how we can bring about this last possibility. Positive thinking alone won't get us there.

SORRY TO BOTHER YOU

BRIAN CHRISTIAN
Author, *The Most Human Human: What Artificial Intelligence Teaches Us About Being Alive*

Before the written word, when we wanted something we had no choice but to ask someone. Growing up, I remember asking my mother on numerous occasions what some unfamiliar word meant, and she would unfailingly reply, tongue only partly in cheek, "What do I look like, a dictionary?" I don't think she intended to become an allegory for AI, but she did instill in me some dimly understood sense that it was rude to ask of a flesh-and-blood human being what could just as easily be asked of an artifact.

That was some decades ago. In the present, we've subconsciously internalized as well as extended this principle. When we stop someone to ask for directions, there's usually an explicit or implicit "I'm sorry to bring you down to the level of Google temporarily, but my phone is dead, see, and I require a fact." It's a breach of etiquette, on a spectrum with asking someone to temporarily serve as a paperweight or a shelf.

I've seen this breach, also, in brief conversational moments when someone asks a question of someone else—a number, a date, a surname, the kind of question you could imagine being on a quiz show, some obscure point of fact—and the other person grimaces or waves off the query. They're saying, "I don't know. You have a phone, don't you? You have the entire Internet, and you're disrespecting me, wasting my time, using me."

Not for nothing do we now have the sarcastic catchphrase, "Here, let me Google that for you."

As things stand, there are still a few arenas in which only a human brain will do the trick—in which the relevant information and experience lives only in humans' brains and so we have no choice but to trouble those brains when we want something. "How do those latest figures look to you?" "Do you think Smith is bluffing?" "Will Kate like this necklace?" "Does this make me look fat?" "What are the odds...?" These types of questions may well offend in the twenty-second century. They require a mind—but *any* mind will do, and so we reach for the nearest one.

There's a memorable scene in the 1989 romantic comedy *Say Anything*, where Ione Skye returns apologetically to John Cusack, professing her love and asking for his forgiveness. "One question," he says. "You're here 'cause you need *someone* or 'cause you need *me*?" When artifacts can say anything requiring general intelligence, this will be the question repeated underneath every human interaction like a hidden mantra, the standard to which all engagement will be subjected.

When we human beings leave the movie theater or the playhouse or the museum, the question on all of our lips is, "What did you think?" This question will be one of the few to outlast the coming of AI. We'll simply take care to italicize the *you* rather than the *think*.

MORAL MACHINES

BENJAMIN K. BERGEN

Associate professor, cognitive science, UC San Diego; author, *Louder Than Words: The New Science of How the Mind Makes Meaning*

Machines make decisions for us. A trading machine in Manhattan detects a change in stock prices and decides in microseconds to buy millions of shares of a tech company. A driving machine in California detects a pedestrian and decides to turn the wheels to the left.

Whether these machines are "thinking" or not isn't the issue. The real issue is the decisions we're empowering them to make. More and more, these are consequential. People's savings depend on them. So do their lives. And as machines begin to make decisions that are more consequential for humans, for animals, for the environment, and for national economies, the stakes get higher.

Consider this scenario: A self-driving car detects a pedestrian running out in front of it across a major road. It quickly apprehends that there's no harm-free course of action. Remaining on course would cause a collision and inevitable harm to the pedestrian. Braking quickly would cause the car to be rear-ended, with the attendant damage and possible injuries. So would veering off-course. What protocol should a machine use to decide? How should it quantify and weigh different types of potential harm to different actors? How many injuries of what likelihood and severity are worth a fatality? How much property damage is worth a 20 percent chance of whiplash?

Questions like these are hard to answer. They're questions you can't solve with more data or more computing power. They're about what's morally right. We're charging machines with moral decisions. Faced with a conundrum like this, we often turn to humans as a model. What would a person do? Let's re-create that in the machine.

The problem is that when it comes to moral decisions, humans are consistently inconsistent. What people say they believe is right and what they actually do often don't match (recall the case of Kitty Genovese). Moral calculus differs over time and from culture to culture. And the details of each scenario affect people's decisions: Is the pedestrian a child or an adult? Does the pedestrian look intoxicated? Does he look like a fleeing criminal? Is the car behind me tailgating?

What's the right thing for a machine to do?

What's the right thing for a human to do?

Science is ill-equipped to answer moral questions. Yet the decisions we've already handed to machines guarantee that someone will have to answer them, and there may be a limited window left to ensure that that someone is human.

AFTER THE PLUG IS PULLED

LAURENCE C. SMITH
Professor and chair of geography;
professor of Earth, Planetary, and Space Sciences,
UCLA; author, *The World in 2050*

What's the big deal about machines that think? For a small group of philosophers and theologians, I get it—but for the rest of us, artificial intelligence will just be the latest incremental step in a long stampede of technological encroachment that has already changed the world almost beyond recognition.

For that very important job of thinking that seeks to solve problems, there's little doubt that adaptive, machine-based learning will do better than any one human brain (or even an entire conference of experts). Machines already think more deeply about your consumer preferences than you do, through creepy, financially motivated adaptive algorithms that track your online behavior. But other purposes now underway include smarter policing and identifying potential child-abuse situations, both drawn from disparate data pulled together to identify a broader pattern.

That process has been a hallmark of human thinking since we walked out onto the savannah, and as the world's problems become more dire and more complicated, we ought to accept any effective tool to battle them. I could live with a partnership with machine learning in order to make complex modern life more resource-efficient in a way that human brains cannot. A world of sustainably grown food, sufficient clean water for humans and ecosystems, and comfortable, energy-efficient lodging is still possible and could be advanced in part by thinking machines.

History suggests that the partnership will proceed in an incremental way, relatively unnoticed by busy people living out their busy lives. But for the sake of argument, let's assume that our worst fears come true, things get out of hand, and at some point thinking machines topple the reign of *Homo sapiens*. Then what? I have no doubt we'd somehow manage to pull the plug. A great retoppling would occur, and we'd once again regain dominion over the lands, oceans, and skies. Depending on the depth of the integration and the height of the fall, the human experience might even revert to something more closely resembling the world of ten millennia ago than of today, as we relearn from scratch the basics of acquiring food, water, shelter, and transport without the help of our thinking machines.

MONITORING AND MANAGING THE PLANET

GIULIO BOCCALETTI

Physicist; atmospheric and oceanic scientist;
global managing director for water, Nature Conservancy

In 1922, the mathematician Lewis Fry Richardson imagined a large hall full of "computers"—people who, one hand-calculation at a time, would advance numerical weather prediction. Less than 100 years later, machines have improved the productivity of that particular task by up to fifteen orders of magnitude, with the ability to process almost a million billion similar calculations per second.

Consider the growth in heavy-labor productivity by comparison. In 2014, the world used about 500 exajoules—a billion billion joules—of primary energy to produce electricity and fuel manufacturing, transport, and heat. Even if we posit that all of that energy went into carrying out physical tasks in aid of the roughly 3 billion members of the global labor force (and it didn't), assuming an average adult diet of 2,000 calories per day would imply roughly 50 "energy laborers" for every human. More stringent assumptions would still lead to, at most, an increase of a few orders of magnitude in effective productivity of manual labor.

We've been wildly successful at accelerating our ability to think and process information, more so than for any other human activity. The promise of AI is to deliver another leap in increasing the productivity of specific cognitive functions—

ones where the sophistication of the task is also orders of magnitude higher than previously possible.

Keynes would probably have argued that such an increase should ultimately lead to a fully employed society, with more free time and a higher quality of life for all. The skeptic might be forgiven for considering this a case of hope over experience. Whereas there's no question that specific individuals will benefit enormously from delegating tasks to machines, the promise of greater idleness from automation has yet to be realized, as any modern employee—virtually handcuffed to a portable device—can attest.

So if we're going to work more, deeper, and with greater effectiveness thanks to thinking machines, choosing wisely what they'll be "thinking" about is important. Indeed, it would be a shame to develop all this intelligence and then spend it on thinking really hard about things that don't matter. As ever in science, selecting problems worth solving is a harder task than figuring out how to solve them.

One area where the convergence of need, urgency, and opportunity is great is in the monitoring and management of our planetary resources. Despite the dramatic increase in cognitive and labor productivity, we haven't fundamentally changed our relationship to Earth: We're still stripping it of its resources in order to manufacture goods that become waste relatively quickly, with essentially zero end-of-life value to us. A linear economy on a finite planet, with 7 billion people aspiring to become consumers. Our relationship to the planet is arguably more productive but not much more intelligent than it was 100 years ago.

Understanding what the planet is doing in response and managing our behavior accordingly is a complicated problem, whose solution is hindered by colossal amounts of imperfect information. From climate change to water availability to the management of ocean resources to the interactions between ecosystems and working landscapes, our computational approaches are often inadequate to conduct the exploratory analyses required to understand what's happening, to process the exponentially growing amount of data about the world we inhabit, and to generate and test theories of how we might do things differently.

We have almost 7 billion thinking machines on this planet already, but for the most part they don't seem terribly concerned with how sustainable their life on this planet actually is. Few can see the whole picture in ways that make sense to them, and those who do are often limited in their ability to respond. Adding cognitive capacity to figure out how we fundamentally alter our relationship with the planet is a problem worth thinking about.

PANEXPERIENTIALISM

IAN BOGOST
Ivan Allen College Distinguished Chair in Media Studies and Professor of Interactive Computing, Georgia Institute of Technology; founding partner, Persuasive Games LLC; author, *Alien Phenomenology, or What It's Like to Be a Thing*

The Search for Extra Terrestrial Intelligence (SETI) names the globally distributed projects, people, and institutions searching the cosmos for signs of intelligent life. SETI's methods mostly entail scanning for the emission of electromagnetic radiation, an exhaust that's assumed to emanate from civilizations with advanced technologies.

Like the quest to build intelligent machines, the search for intelligent aliens makes assumptions about what intelligence is and what aliens are. SETI assumes that alien life is intelligent if it matches humans' science-fictional expectations for intelligence: animalian creatures with communication devices and spaceships and the like.

Critics of SETI sometimes invoke what are called uniformitarian objections. Uniformitarianism names the assumption that the same conditions and laws apply everywhere throughout time and space. SETI is uniformitarian in its assumption that all alien intelligence would be the same—namely, like human intelligence (but smarter, of course). But it's just as compelling to think otherwise. The philosopher Nicholas Rescher, for example, has observed that if there's intelligence in the universe, we might not be able to identify it as intelligence. True alien intelligence would differ from ours not only in its cosmic location but also in its very nature. As Doris and David Jonas put it some forty years ago, different sensory capacities produce different "slits" for perceiving, explaining, and interacting with reality.[12]

This means that alienness is not just "out there" but all around us. You might find your cat to be intelligent in a certain way, or your smartphone, or your car, or a hypothetical future robot, or, given the right perspective, even your houseplant or your toaster.

The dream of thinking machines is really no different than the dream of intelligent aliens. It just replaces the biological, cosmic, entropy-fashioned alien of afar with the mechanico-electronic, human-fashioned machine in our midst. And if SETI and its kin make a uniformitarian mistake in the cosmos, efforts to theorize and create artificial intelligence and thinking machines make the same mistake here on Earth.

Perhaps the best evidence for thinking machines' reliance on the particular mode of intelligence that humans experience can be found in our fictional doomsday worst-case scenarios for AI. The fear of a robot or computer apocalypse of the *Terminator* or *Berserker* or *Matrix* varieties depends on machine intelligence besting humans to the point where it realizes that the best option is to destroy and replace humans (or, in the Kurzweilian Singularity version of AI fantasy, humans willingly submit to their computer overlords in order to achieve immortality). Closer to home than to doomsday, our fear of machine intelligence also expresses itself in a concern over the role of human thought and labor in an economy increasingly run by mechanical and electronic machines.

This is one vision of thinking machines, but it needn't be the only one. Thinking about thinking machines turns out to be so narrow and anthropocentric it's surprising we haven't given up on it out of boredom rather than on contra-uniformitarian grounds. Instead of asking if machines can think, or what we need to do to cause them to think, or how we'd know if they were thinking, what if we just assumed that all "machines" did something akin to thinking and then we attempted to characterize what "thinking" might mean?

In philosophy, there are already directions toward such an approach. Contrary to the emergentist position that most AI advocates hold—that mind emerges from specific material conditions, whether in biological or computational entities—panpsychists claim that "minds" are everywhere, in some sense. Panpsychism bears some relationship to certain Buddhist doctrines that encourage awareness of the animism in nature. But panpsychism risks the same erroneous uniformitarianism as SETI or AI—namely, that a mind akin to that of a human (or at least of an animal) is the model for all other minds. A more promising philosophical position is that of panexperientialism, the position that everything has something like experience, even if it might be very different from that of a human being.

When we think about thinking machines, we usually think of a particular sort of machine and a particular sort of thinking—electronic and (super)human, respectively. But what if instead we allowed for the possibility that we've been missing out on all the "thinking" being done by all the other kinds of machines surrounding us—toasters, garage doors, automobiles? This may seem like a ludicrous waste of time, but it doesn't take long to prove useful. If the purpose of thinking about machines like AIs and robots and computers is in part to struggle with the question of what living with them as neighbors and companions and even citizens might look like, then we ought to start by taking more seriously all the machines already around us that could be said to have taken on those roles and which we nevertheless ignore.

WHEN IS A MINION NOT A MINION?

AUBREY DE GREY
Gerontologist; chief science officer, SENS Foundation; author, *Ending Aging*

If asked to rank humanity's problems by severity, I would give the silver medal to the need to spend so much time doing things that give us no fulfillment—work, in a word. I consider that the ultimate goal of artificial intelligence is to hand off this burden, to robots that have enough common sense to perform those tasks with minimal supervision.

But some AI researchers have altogether loftier aspirations for future machines. They foresee computer functionality that vastly exceeds our own in every sphere of cognition. Such machines would not only do things that people prefer not to, they would also discover how to do things that no one can yet do. This process can, in principle, iterate: The more such machines can do, the more they can discover.

What's not to like about that? Why do I not view it as a superior research goal to that of building machines with common sense (which I'll call "minions")?

First, there is the well-publicized concern that such machines might run amok—especially if the growth of a machine's skill set (its "self-improvement") is not iterative but recursive. What researchers mean by this is that enhancements might be not only to the database of things a machine can do but to its algorithms for deciding what to do. Some have suggested that this recursive self-improvement might be exponential (or faster), creating functionality we cannot remotely comprehend, before we can stop the process. So far so majestic, if it weren't for the idea that the trajectory of improvement would itself be out of our control, such that these superintelligent machines might gravitate to "goals" (metrics by which they decide what to do) that we dislike. Much work has been done on ways to avoid this "goal creep" and to create a reliably, permanently "friendly," recursively self-improving system—but with precious little progress.

My reason for believing that recursive self-improvement is not the right ultimate goal for AI research is not the risk of unfriendly AI, though; rather, it is that I strongly suspect that recursive self-improvement is mathematically impossible. In analogy with the so-called halting problem of determining whether or not a program terminates, I suspect there's a yet-to-be-discovered measure of complexity by which no program can ever write another program (including a version of itself) that is an improvement.

The program written may be constrained to be, in a precisely quantifiable sense, simpler than the program that does the writing. It's true that programs can draw on the outside world for information on how to improve themselves—but I claim (a) that that only delivers far less scary iterative self-improvement rather than recursive, and (b) that anyway it will be inherently self-limiting, since once these machines become as smart as humanity, they won't have any new information to learn. This argument isn't anywhere near ironclad enough to give true reassurance, I know, and I bemoan the fact that (to my knowledge) no one is really working to seek such a measure of depth or prove that none can exist. But it's a start.

In contrast, I absolutely am worried about the other reason why I stick to the creation of minions as AI's natural goal. It is that any creative machine—whether technologically, artistically, whatever—undermines the distinction between man and machine. Humanity has enormous uncertainty already regarding what rights various nonhuman species have. Since objective moral judgments build on agreed norms, which themselves arise from inspection of what we would want for ourselves, it seems impossible even in principle to form such judgments concerning entities that differ far more from us than animals do from one another. So I say we shouldn't put ourselves in the position of needing to try. For illustration, consider the right to reproduce despite resource limitations. Economic incentive-based compromise solutions seem to work adequately. But how can we identify such compromises for "species" with virtually unlimited reproductive potential?

I contend that the possession of common sense does not engender these problems. I define common sense, for present purposes, as the ability to process highly incomplete information so as to identify a reasonably close-to-optimal method for achieving a specified goal, chosen from a parametrically prespecified set of alternative methods. This explicitly excludes the option of "thinking"—of seeking new methods, outside the prespecified set, that might outperform anything within the set.

Thus, again for illustration, if the goal is one that should ideally be achieved quickly, and can be achieved faster by many machines than by one, the machine won't explore the option of first building a copy of itself unless that option is prespecified as admissible, however well it may "know" that doing so would be a good idea. Since admissibility is specified by inclusion rather than exclusion, the risk of "method creep" can (I claim) be safely eliminated. Vitally, it's possible to prevent recursive self-improvement (if it turns out to be possible after all!) entirely.

The availability of an open-ended vista of admissible ways to achieve one's goals constitutes a good operational definition of "awareness" of those goals. Awareness implies the ability to reflect on the goal and on one's options for achieving it, which amounts to considering whether there are options one hadn't thought of.

I could end with a simple "So let's not create aware machines," but any possible technology that anyone thinks is desirable will eventually be developed, so it's not that simple. What I say instead is, let's think hard now about the rights of thinking machines, so that well before recursive self-improvement arrives we can test our conclusions in the real world with machines that are only slightly aware of their goals. If, as I predict, we thereby discover that our best effort at such ethics fails utterly even at that early stage, maybe such work will cease.

NOT BUGGY ENOUGH

MICHAEL I. NORTON
Professor of business administration, Harvard Business School; coauthor (with Elizabeth Dunn), *Happy Money: The Science of Smarter Spending*

A pervasive human fear emerged in the twentieth century, one that grows stronger with each new doomsday prediction: Inevitably, as artificial intelligence advances, some unforeseen computer bug will cause computers to revolt and take over the world.

My concern is the opposite: that as artificial intelligence advances, it will not be buggy enough. Thinking machines that are perfectly self-correcting, self-optimizing, and self-perfecting, so that the square peg always ends perfectly in the square hole, will also be machines that fail to inculcate the random sparks of insight coming from the human tendency to be buggy—to try to fit square pegs into round holes, or, more broadly speaking, to notice the accidental but powerful insights that can arise as a by-product of solving a shape/hole problem.

Consider the power of noticing. The reason we can enjoy macaroni and cheese in a matter of seconds? While working at Raytheon, Percy Spencer walked in front of a machine and happened to notice that his chocolate bar had melted. Why? The machine generated microwaves. Instead of trying to optimize the magnetron to avoid future chocolate failure, Spencer had a flash of insight that the melted chocolate was the harbinger of something bigger.

Consider the power of accidents. Rubber was doomed to specialized usage because of its failure to withstand extreme temperatures—until Charles Goodyear slipped up and dropped some rubber on a hot stove. Instead of taking steps to ensure no future mistakes, Goodyear noticed something interesting, and the result was vulcanized, weatherproof rubber.

Finally, consider the power of human "bugs"—our biases. For example, optimism makes us believe we can get to the moon, cure all diseases, and start a successful business in a terrifying location whose previous tenant fled. The endowment effect causes us to overvalue what we have, what we ideate, and what we create—even when no one else agrees. But is abandoning all endeavors at the first sign of failure and pursuing one that seems more successful always optimal? The dogged scientists (think of Galileo and Darwin) who persist in the face of generally accepted explanations are being stubborn—being buggy—but the result can be genius. In large part, it's the bugs that make us—and any form of intelligence—human.

MORE FUNK, MORE SOUL, MORE POETRY AND ART

THOMAS A. BASS
Professor of literature, University at Albany, SUNY;
author, *The Spy Who Loved Us*

Thinking is good. Understanding is better. Creating is best. We're surrounded by increasingly thoughtful machines. The problem lies in their mundanity. They think about landing airplanes and selling me stuff. They think about surveillance and censorship. Their thinking is simpleminded, if not nefarious. Last year a computer was reported to have passed the Turing Test. But it passed as a thirteen-year-old boy, which is about right, considering the preoccupations of our jejune machines.

I can't wait for our machines to grow up, to get more poetry and humor. This should be the art project of the century, funded by governments, foundations, universities, businesses. Everybody has a vested interest in getting our thinking more thoughtful, improving our understanding, and generating new ideas. We've made a lot of dumb decisions lately, based on poor information or too much information or the inability to understand what this information means.

We have numerous problems to confront and solutions to find. Let's start thinking. Let's start creating. Let's agitate for more funk, more soul, more poetry and art. Let's dial back on the surveillance and sales. We need more artist-programmers and artistic programming. It's time for our thinking machines to grow out of an adolescence that has lasted now for sixty years.

THE FUTURE IS BLOCKED TO US

HANS ULRICH OBRIST
Codirector of Exhibitions and Programmes, director of International Projects, Serpentine Galleries, London; author, *Ways of Curating*

In his poem of the same name (which also serves as the title to Adam Curtis's seminal documentary), Richard Brautigan portends a future "all watched over by machines of loving grace" or, by implication, "thinking" machines. In what follows, I use the term *thinking* to refer to machines that think on purely algorithmic and computational lines—machines coded by engineers rather than machines that might, or could be, truly sentient.

Curtis argues that we're living in a "static culture," a culture too often obsessed with sampling and recycling the past. He implies that the Age of the Thinking Machine is resulting in ossification rather than renewal. As our lives become increasingly recorded, archived, and accessed, we have become cannibals driven to consume our history and terrified of transgressing its established norms.

To some extent, the future is blocked to us; we're stuck in stasis; we're stuck with a version of ourselves that's becoming increasingly narrow. No thanks to recent tools such as "recommender systems," we're lodged in a seemingly endless feedback loop of "If you liked that, you'll love this." As we might become increasingly stuck in Curtis's idea of the "you-loop," so the nature of what it means to be human might be compromised by job-hogging machines that will render many of us obsolete. This *Edge* Question points to the next chapter in human history/evolution; we're facing the beginning of a new definition of man, a new civilization.

An optimistic approach to the question of machines that think comes from the Lebanese American poet Etel Adnan, who celebrates her ninetieth birthday this year. For her, thinking machines may think better than we do—to start with, because they won't tire as fast as we do. They may also ask questions we're not used to answering. Etel has said that what shook her most was of another order. Seeing a picture of a robot, a life-sized structure resembling the metal armory of a medieval knight, she immediately envisioned an old woman (or an old man) utterly alone, as so many of them are nowadays, having for sole companionship such a creature-like object, capable of doing things, of talking, and the old person falling in love with it, which made her cry.

The idea of machines that think plays a role in the work of another artist, Philippe Parreno, who works with algorithms, which for him have replaced cinema as a model of the perception of time. In the last century, Gilles Deleuze's writings on repetition and difference in cinema emphasized that film unfolds in time and is composed of ever-differentiating planes of movements. As Parreno shows, Deleuze transposed those theories to discuss the mechanized and standardized movements of film as a means of reproducing or representing life. Parreno's work with machines that think explores how, today, algorithms are changing our relation to movements, rhythms, and durations—or, to put it in Leibniz's terms, the question will be, "Are machines spiritual automatons?"

AN IMMATERIAL THINKABLE MACHINE

KOO JEONG-A
Conceptual artist

We can step up to immaterial science, then make an immaterial thinkable machine by starting a very simple manual to program.

BAFFLED AND OBSESSED

RICHARD FOREMAN
Playwright and director;
founder, Ontological-Hysteric Theater

Perhaps the question (a question being a problem) is really a false problem? Obviously machines calculate, "write" poems, organize vast amounts of material, etc. Is any of this "thinking"? (Do I think? But what do I really do when I think I'm thinking?) One answer: I experience anguish—a "hole" in my "inherited" smoothly proceeding discourse (inner or outer).

So—different possibilities?

I *fall* into that hole—i.e., either I'm so baffled I stop thinking or I come up from its emptiness with an idea or solution (in my case, work of art) that obtains a so-called desired result: i.e., others (some others) react in some fashion. Not very interesting, really. And is this what "result-oriented" machines do? That's what I think—baffled, and obsessed. (And are machines ever baffled? Do they ever "stop thinking" when thinking?)

I know when I edit film, my Final-Cut software can crash when the machine gets somehow overloaded, but this crash doesn't create a hole in the machine with the resultant possibility of an emptiness that "feeds." (When *I* "crash," something may enter my dim, nonfocused consciousness, and I may go in a new direction.) This is part of my thinking that I don't think a machine can do. (Am I wrong? I have no vast knowledge of machines.) I'm stupid, so I flail about and hit something sometimes—deep and wonderful? One chance in a hundred—maybe.

OK. I experience a hole that I'm conditioned to believe should be filled (with the already known, usually). I maintain that to fill it is to die a little bit. Better—what I *can* do is build a shrine around it that makes the hole ever more "resonant" while still "empty." (I suggest this is how the serious artist may work—plus who else?) But "building" around the hole is not creative thinking, it's what can be done in place of creative thinking—though it does make something "to think about." But the hole is the point: the evocation and amplification of "mystery," which echoes the "big mystery" that I "think" real "thinking" is about. (Does that confine me in the tight box of "being an artist?"—i.e., irresponsible?)

OK, machines can "sort of" think with ever greater degrees of power and complexity, spinning wider and wider webs, but the web is never a *single hole*. Machines think? A tautology. They do facilitate my living and functioning in society. Obviously one kind of thinking—but not the mysterious going in circles on circles, producing the sparks of friction that are "the essence" (dare I say that?).

I'm worried. Can I answer the question "What do you think about machines that think?" Worried, yes, but machines can't worry (can they?). OK—to "worry," meaning the inability to think of anything else, unable to get off the very *spot* of worry. Result—*blackout!* Draw a blank. But in that *blank* I go off on a possibly productive (but to what *end*, and must there be one?) tangent. Can a machine go off on a tangent? Would that be thinking?

What we normally call "thinking" is obsessively goal-oriented. But is there a kind of goal serving no purpose—and can only a human brain latch on to such a perverse idea? Which could lead who knows where? OK—obvious by now—why did I have to go in circles to make the perhaps obvious point that, to my mind, machines that think are the contemporary Trojan Horse. Everyone (me included) wants the many sweets they offer, while those very sweets do *mold* us in their image, thereby smothering the blankness of deep creativity inside each of us. And why did I have to go in circles to get *here*, where I'm offering an opinion—worth not nearly as much as the *rhythm* of my circling . . . a Hole. Yes, I am caught in a trap of my own making—just like everyone. But *not* like machines that think! The trap *they're* in—well, they cannot "know."

WHO'S AFRAID OF ARTIFICIAL INTELLIGENCE?

RICHARD H. THALER

Father of behavioral economics; director, Center for Decision Research, University of Chicago Booth School of Business; author, *Misbehaving: The Making of Behavior Economics*

My brief remarks on this question are framed by two one-liners that happen to have been uttered by brilliant Israelis. The first comes from my friend, colleague, and mentor Amos Tversky. When asked once what he thought about artificial intelligence, Amos quipped that he didn't know much about it, his specialty was natural stupidity. (Before anyone gets on their high horse, Amos didn't actually think people were stupid.)

The second joke comes from Abba Eban, who was best known in the United States when he served as Israel's ambassador to the United Nations. Eban was once asked if he thought Israel would switch to a five-day workweek. Nominally, the Israeli workweek starts on Sunday morning and goes through midday Friday, though a considerable amount of the "work" done during those five-and-a-half days appears to take place in coffeehouses. Eban's reply to the query was, "One step at a time. First, let's start with four days, and go from there."

These jokes capture much of what I think about the risks of machines taking over important societal functions and then running amok. Like Tversky, I know more about natural stupidity than about artificial intelligence, so I have no basis for forming an opinion about whether machines can think, and if so, whether such thoughts would be dangerous to humans. I leave that debate to others. Like anyone who follows financial markets, I'm aware of incidents such as the Flash Crash in 2010, when poorly designed trading algorithms caused stock prices to fall suddenly, only to recover a few minutes later. But this example is more an illustration of artificial stupidity than hyperintelligence. As long as humans continue to write programs, we'll run the risk that some important safeguard has been omitted. So, yes, computers can screw things up, just like humans with "fat fingers" can accidently issue an erroneous buy or sell order for gigantic amounts of money.

Nevertheless, fears about computers taking over the world are premature. More disturbing to me is the stubborn reluctance in many segments of society to allow computers to take over tasks that simple models perform demonstrably better than humans. A literature pioneered by psychologists such as the late Robyn Dawes finds that virtually any routine decision-making task—detecting fraud, assessing the severity of a tumor, hiring employees—is done better by a simple statistical model than by a leading expert in the field. Let me offer just two illustrative examples, one from human-resource management and the other from the world of sports.

First, let's consider the embarrassing ubiquity of job interviews as an important, often the *most* important, determinant of who gets hired. At the University of Chicago Booth School of Business, where I teach, recruiters devote endless hours to interviewing students on campus for potential jobs—a process that selects the few who will be invited to visit the employer, where they will undergo another extensive set of interviews. Yet research shows that interviews are nearly useless in predicting whether a job prospect will perform well on the job.

Compared to a statistical model based on objective measures such as grades in courses relevant to the job in question, interviews primarily add noise and introduce the potential for prejudice. (Statistical models don't favor any particular alma mater or ethnic background and cannot detect good looks.)

These facts have been known for more than four decades, but hiring practices have barely budged. The reason is simple: Each of us just *knows* that if *we* are the one conducting an interview, we will learn a lot about the candidate. It might well be that *other* people are not good at this task, but I am! This illusion, in direct contradiction to empirical research, means that we continue to choose employees the same way we always did. We size them up, eye to eye.

One domain where some progress has been made in adopting a more scientific approach to job-candidate selection is sports, as documented by the Michael Lewis book and movie *Moneyball*. However, it would be a mistake to think there has been a revolution in how decisions are made in sports. It's true that most professional sports teams now hire data analysts to help them evaluate potential players, improve training techniques, and devise strategies. But the final decisions about which players to draft or sign, and whom to play, are still made by coaches and general managers, who tend to put more faith in their gut than in the resident geek.

An example comes from American football. David Romer, an economics professor at Berkeley, published a paper in 2006 showing that teams choose to punt far too often, rather then trying to "go for it" and get a first down, or score.[13] Since the publication of his paper, his analysis has been replicated and extended with much more data, and the conclusions have been confirmed. The *New York Times* even offers an online "bot" that calculates the optimal strategy every time a team faces a fourth-down situation.

But have coaches caught on? Not at all. Since Romer's paper was published, the frequency of going for it on fourth down has been flat. Coaches, who are hired by owners based in part on interviews, still make decisions the way they always have.

So pardon me if I don't lose sleep worrying about computers taking over the world. Let's take it one step at a time, and see if people are willing to trust them to make the easy decisions at which they're already better than humans.

I SEE A SYMBIOSIS DEVELOPING

SCOTT DRAVES
Software artist; creator, Electric Sheep

I think thinking about machines that think is the most interesting thing to think about. Why? Because the possible implications of this phenomenon are profound. Cosmic even.

"Thinking machines" have been with us for a long time. There are two ways to understand this, depending on which word you start with. Let's start with *machines* first, and by that these days we really mean computers. Computers started out, well, pretty mechanical. But they keep getting more and more subtle. Even the computers of the 1980s could perform some remarkable feats with expert systems and databases. Today we've passed the point where we could explain in detail how voice recognition and natural language allow our phones to answer a question spoken by a child. "Magical" is hardly hyperbole. But is this really thinking? Not yet, but it's a good start, and the trend is accelerating. True, the goal still seems far away. Instead of considering our climb step by step, look up and consider what lies at the summit. Can anything halt our progress?

Certainly the future of chip technology is in doubt. Moore's Law has been good to us, and has dodged a few bullets, but it's ending. Historically, new technologies have appeared just in time to keep the exponential growth of computation on schedule, but this is no given. Perhaps the next leap is incredibly difficult and will take fifty years. Or it may never happen, though we can always add more chips in parallel. The schedule is an interesting question, but it pales in comparison to pondering the destination.

Now let's take up the word *think*. The other way that thinking machines have been around for a long time is ourselves. Biological brains have been thinking for millions of years. A brain follows the laws of physics, which are a mechanical set of equations. In principle, a good physics simulator could, very slowly, simulate a brain and its environment. Surely this virtual brain would be a machine that thinks.

The remaining question is, How much physics is required to make the simulation work? Would classical physics, electricity, and chemistry do? Would quantum logic (or beyond) be required? The consensus is strongly in favor of the idea that classical physics suffices (the Emperor's New Mind has been rejected). Hence I think of my brain and body as a giant machine made up of an octillion molecules: many, many tiny magnetic Tinkertoys whose behavior is well understood and can be simulated. There are good reasons to believe a statistical approximation of physics can provide the same results. But again, this affects only the schedule, not the destination. The important question is, How do thinking and consciousness emerge from this complex machine? Is there some construction, some bridge, from the digital and virtual to the analog, organic, and real?

These threads meet with the merger of human and computer substrates. Smartphones are rapidly becoming indispensable parts of ourselves. The establishment has always questioned the arrival of new media, but adoption of these extensions of ourselves continues apace. A lot of ink has been spilled over the coming conflict between human and computer, be it economic doom, with jobs lost to automation, or military dystopia teeming with drones. Instead, I see a symbiosis developing. And historically, when a new stage of evolution appeared— like eukaryotic cells, or multicellular organisms, or brains—the old system remained and the new system worked with it, not in place of it.

This is cause for optimism. If digital computers are an alternative substrate for thinking and consciousness, and digital technology is growing exponentially, then we face an explosion of thinking and awareness. This is a wave we can ride, but doing so requires us to accept the machine as part of ourselves, to dispense with pride and recognize our shared essence. Essentially we must meet change with love instead of fear. I believe we can.

REIMAGINING THE SELF IN A DISTRIBUTED WORLD

MATTHEW RITCHIE
Artist

Will it happen? It already has. With the gradual fusion of information-storing-and-reporting technologies at the atomic and molecular scales, and the scaling up of distributed and connected information-storing-and-reporting devices at the social and planetary scale (which already exceeds the number of human beings on the planet), the definitions of both *machine* and *thinking* have shifted to embrace both inorganic and organic "complexes" and "systemic decisions" as interchangeable terms—mechanically, biologically, physically, intellectually, and even theologically.

Near-future developments in biotechnology and transhuman algorithmic prediction systems will quickly render many of the last philosophical distinctions between *observing*, *thinking*, and *deciding* obsolete, and quantitative arguments meaningless. Once those barriers are crossed and the difference between "a machine that thinks" and "a biological system that thinks" becomes trivial, the focus immediately shifts to qualitative questions—human definitions of *intentionality* and *agency* for thinking machines.

What will that mean for us? Does the existence of thinking machines—whether arranged in an inorganic or quantum array or a biochemical holarchy—diminish human agency or extend it? Are we willing to extend our definition of ourselves not just to authored and mechanical systems but to the independent and symbiotic systems that already inhabit us—the trillions of bacteria in our gut (which alter our mental states by manipulating chemical pathways) and the biochemical trackers, agents, and augmentals we ingest? What will it mean to fully extend ourselves into and through thinking machines?

An AI will quickly find its way to the world library, the Web. Once there, it will join the many quasi-human systems, distributed crowd intelligences, and aggregated thinking machines already inhabiting this space and will quickly learn to generate or simulate the models of continuous conscious reflectivity and mirror selves found there and easily reproduce or co-opt the apparently complex alternative identities and ambiguities that define the Web.

Drawing distinctions between the real and unreal for an independent, evolving, functional, intelligent system will be the most significant discussion of all. How will it be taught? In object-oriented ontology, the universe is presented as already being full of objects and qualities, which are constituted into meaningful systems by human consciousness. Just what are the qualitative differences between spontaneously created thinking systems—or composites of objects and qualities—and artificially created thinking systems? What will happen if or when one of these rejects or surpasses the essential philosophies of its makers?

Redefining the nature and role of the human thinking self, as a self-othering, self-authoring, and self-doctoring system whose precise nature and responsibilities have been argued about since the Enlightenment, will be a critical question, linked to questions of shared community and our willingness to address the ethical determination and limits of independent systems whose real-world consequences cannot be ignored.

Are such systems alive? What are their rights and responsibilities? Since the Supreme Court decisions that elevated corporations to the status of individuals, we've accepted the legal precedent that nonhuman aggregated "thinking machines" can be an integral part of our political and cultural life and struggled with how to restrain nonhuman systems in human terms. It will be no small task to integrate the complex and diverse human ethical, creative, and representational belief systems into a meaningful civic process that defines an ability to think as a basis for citizenship.

The weakest counterargument against the thinkinghood of artificial life, often coming from the humanities, is a vaguely medieval, mystical assertion that human perceptions of symmetry and beauty can never be matched by machines. It's an article of faith in the interpretive arts that a machine can never do a human being's work—but it's just a comforting illusion to suppose that the modest aesthetic standards of any given contemporary taste cannot be codified and simulated. Machines already perform bestselling pop songs and take spectacular photographs of other planets and stars. There are already video games as beautiful as films. Whether a thinking machine can learn how to write a symphony or sketch a masterpiece is only a question of time. Perhaps a more significant question is whether it can learn how to make a great work of art, ultimately achieving through sheer capacity what no human could through improvisation. Part of the enormously larger and newly horizontal distributed network of cultural practice, supported by new technologies, has indeed begun to fall into what Jaron Lanier has described as "hive thinking," supporting the gloomiest cultural predictions.[14] But as Heidegger proposed, the danger of unexamined scientific rationalism is that the most reductive definition of *object* as "machine" or "system" can be extended to the universal scale in every sense, becoming a self-justifying and ethically vacant rationale for the mechanization of the self. The ensuing fantasies—Samuel Butler's vital machines, H. G. Wells's shadowy world of make-work, or the fear of becoming components in a supersystem or matrix—are primarily failures of human imagination.

The emergence and definition of new kinds of dynamically aggregated "information citizens," and aggregated working platforms whether collective or individual, biological, corporate, national, or transnational, present us with a vast new opportunity—not as members of one species, or as specific composites of objects and qualities, but as a new kind of people—co-owners of an information culture, economy, and ecology, who have, as our shared birthright, access to every culture and every system.

Perhaps hybrid-human-object-system thinking machines are already becoming a vast new source of energy for a depleted historical environment. Perhaps we even have an opportunity to redefine the trajectory for artistic practice altogether. Can the time of emergence for thinking machines inspire us to reimagine and redefine what it is to be truly human, to extend ourselves into the infinite? It already has.

IT'S EASY TO PREDICT THE FUTURE

RAPHAEL BOUSSO
Professor of theoretical physics, UC Berkeley

The future, that is, of a simple system with known initial conditions. It's hopeless to make detailed predictions for a complex, poorly understood system like human civilization. Yet a general argument provides some crude but powerful constraints.

The argument is that we're likely to be typical among any collection of intelligent beings. (The collection should be defined by some general criteria we meet, not carefully crafted to make us special.) For example, the probability that a randomly chosen human is among the first 0.1 percent of humans on Earth is, well, 0.1 percent, given no other information. Of course, our ancestors 10,000 years ago would have drawn the wrong conclusion from this reasoning. But among all humans who ever live, 99.9 percent would be correct, so it's a good bet to make. The probability that we're among the first 0.1 percent of intelligent objects, human or artificial, is similarly tiny.

The assignments of probability would have to be updated if, unrealistically, we somehow gained new information proving that human civilization will continue in present numbers for a billion years. This would be one way of finding out that we lost the bet. But we have no such information, so we must assign probabilities accordingly. (This type of reasoning has been articulated by astrophysicists J. Richard Gott and Alexander Vilenkin, among many others.)

The assumption that we may consider ourselves randomly chosen is sometimes questioned, but in fact it lies at the heart of the scientific method. In physics and other sciences, theories almost never predict definite outcomes. Instead, we compute a probability distribution from the theory. Consider a hydrogen atom: the probability of finding the electron a mile from the proton is not exactly zero, just very, very small. Yet when we find an electron, we don't seriously entertain the possibility that it's part of a remote hydrogen atom. More generally, after repeating an experiment enough times to be satisfied that the probability for the outcome was sufficiently small according to some hypothesis, we reject the hypothesis and move on. In doing so, we're betting that we're not highly atypical observers.

An important rule is that we don't get to formulate the question after we make the observation, tailoring it to make the observation look surprising. For example, no matter where we find the electron, in hindsight the probability was small to have found it at that particular spot as opposed to all the other places it could have been. This is irrelevant, as we would have been unlikely to formulate this question before the measurement. Similarly, humans may well be atypical with respect to some variable we've measured: Perhaps most intelligent objects in the visible universe don't have ten fingers. However, our location in the full temporal distribution of all humans on Earth is not known to us. We know how much time has passed, and how many humans have been born since the first humans, but we don't know what fraction of the full time span or of the total number of intelligent observers on Earth this represents. The typicality assumption can be applied to those questions.

Our typicality makes the following two scenarios extremely unlikely: (1) that humans will continue to exist for many millions of years (with or without the help of thinking machines); and (2) that humans will be supplanted by a much longer-lived or much larger civilization of a completely different type, such as thinking machines. If either were true, then we'd be among the very first intelligent observers on Earth, either in time or by number, and hence highly atypical.

Typicality implies our likely demise in the next million years. But it tells us nothing about whether this will come at the hands (or other appendages) of an artificial intelligence; after all, there's no shortage of doomsday scenarios.

Typicality is consistent with the possibility of a considerable number of civilizations that form and expire elsewhere in our galaxy and beyond. By the same reasoning, their duration is unlikely to vastly exceed ours, a tiny fraction of the lifetime of a star. Even if Earthlike planets are common, as observational evidence increasingly suggests, detectable signals from intelligent beings may not be likely to overlap with our own limited attention span. Still, if our interest lies in assessing the predominance of intelligent machines as a final and potentially fatal evolutionary step, the study of distant planetary systems may not be the worst starting point.

FEAR OF A GOD, REDUX

JAMES CROAK
Artist

Artificial intelligence is very fast database searches. The problem with the data is assigning a value to a certain piece of data; how does one value one piece of data more than another? The value would have to be arranged in a trillion value levels to make any sense in which to consider which idea is more important. Since each idea is a combination of many values, the computer would have to design a new algorithm for each part of the equation to perform the combinatorial analysis of the values. Then it would have to design a model to project into the future the outcome of a proposed decision, but since this concept is too difficult for humans to execute and humans would have to design the computer, what are the chances?

Despite these technical barriers to AI, the single most palpable response to its remote possibility is the fear that it will overpower us and treat us badly. They will be better than us and will treat us as we have treated every life-form beneath us—as an evolutionary bridge to our higher life-form. Fear of AI is the latest incarnation of our primal unconscious fear of an all-knowing, all-powerful angry God dominating us—but in a new ethereal form.

Fear of AI also derives from military weapons development, which had the large budgets to steer computer architecture for generations, with its prime directive to fly and find, intercept and destroy. Given the military lineage, we imagine domination, fret that we cannot compete and will become fodder for the next leap of evolution.

But psyche is too chaotic and irrational in its imaginings to ever duplicate in a machine. Could the machine imagine another machine to take over its rote tasks in order to get some rest? If AI appears, will it wonder who its creator is and be faced with the irrationality that sentient organic matter somehow made it? Would it develop a mythology to fill in the gaps? A religion?

What about meaning production, as in the arts? AI shows no ability to free-associate the prevailing philosophy and aesthetic currents into form and thus provide an experience of meaning; it will produce no grand theories to direct society one way or another. This is the largest problem and one not even vaguely addressed in AI: the production of meaning.

Hence the problem with creativity, which a machine cannot have. A machine could have a database of what has been done in the past but cannot free-associate the myriad irrational influences of our inherited and layered brain, with the variations that form from environmental insult in daily living. They can duplicate but not initiate.

TULIPS ON MY ROBOT'S TOMB

ANDRÉS ROEMER
Diplomat, economist, playwright; cofounder
(with Ricardo B. Salinas Pliego), La Ciudad de las Ideas;
coauthor (with Clotaire Rapaille), *Move UP: Why Some Cultures Advance While Others Don't*

To answer the 2015 *Edge* Question, we should start by knowing a little bit about who we are. So let's begin by talking about our most significant organ, the brain. A simplified schema of this complex structure divides it into three parts: the cortex (responsible for rational processes), the limbic (supporting functions including emotion and motivation), and the reptilian (where our most fundamental and primitive drives, survival and reproduction, reside).

The debate about how to think about thinking machines tends to gravitate toward our cortical and limbic brains. The cortex allows us to more accurately assess the cost/benefit that AI carries, regarding things like the relative costs to business of human versus robot labor and the relative value of human versus digital capital, as well as concerns about bioethics, privacy, and national security. It also enables us to plan, attract more and better funding for research and development, and define our public-policy priorities.

In parallel, our limbic brain helps us take precautions and respond with fear or excitement toward the risks or opportunities of developing AI. Here the panacea and the technophobia are immediate emotional reactions. Common fears include being manipulated or replaced by machines; the opportunities include machine expansion of our memory and facilitation of the daily tasks of life.

But we must also be aware of the powerful—even dominant—role of the reptilian brain in our thinking. This means becoming aware of our most primitive responses, our most territorial and emotive way of thinking about the concepts of thinking, machine, robot, intelligence, artificial, natural, and human. The primary preoccupation of the reptilian brain is survival, and although it's not much talked about, the quest for survival is at the heart of our hopes and fears about thinking machines. When we study ancient archetypes, or literature, or the projections in the contemporary debate reflected in the *Edge* Question, a recurrent subconscious instinctive appears: the reptilian binomial, death versus immortality.

Our fear of death is doubtless behind the collective imagining of robots that reproduce and, with their superior thinking, betray and destroy their creators. Such machines seem to pose the most horrifying danger—that of the extinction of everything that matters to us. But our reptilian brains also see in them the savior; we hope that superintelligent machines will offer us eternal life and youth. Intimations of these ways of thinking are embedded in our language. While in English the terms *robot* and *machine* are genderless, the Latin languages, as well as German, differentiate between them: *el robot* is masculine, dangerous, and fearsome, whereas *la máquina* is feminine, protective, and caring.

Jeremy Bentham defined man as a rational being, but we know we're not. All of us sometimes think and act irrationally because of the power of the reptilian brain, and the reptilian drives have been, and remain, at the heart of the evolution of intelligence. Feeling is what is most profound about thinking.

A machine that grows exponentially in speed of data processing every eighteen months, that defeats natural intelligence in a game of chess by sorting through a zillion options move by move, and that can accurately diagnose diseases is highly impressive, but this isn't what it means to think. In order for us to achieve the dream of thinking machines, they'll have to understand and question values, suffer internal conflicts, experience intimacy.

When thinking about machines that think, we should ask ourselves reptilian questions, such as, Would you risk your life for a machine? Would you let a robot be a political leader? Would you be jealous of a machine? Would you pay taxes to ensure a robot's well-being? Would you put tulips on your robot's tomb? Or, more important, Would my robot put tulips on my tomb?

Acknowledging the power of the reptilian in our thinking about machines that think helps us to see more clearly the implications, and nature, of a machine that can genuinely doubt and commit, and the kind of AI we should aspire to. If our biology designed culture as a tool for survival and evolution, nowadays our natural intelligence should lead us to create machines that feel and are instinctual; only then will immortality overcome death.

TOWARD A NATURALISTIC ACCOUNT OF MIND

LEE SMOLIN
Theoretical physicist, the Perimeter Institute; author, *Time Reborn*

"To think" can mean to reason logically, which certainly some machines do, albeit by following algorithms we program into them. Or it can mean "to have a mind," by which we mean that a machine experiences itself as a subject, endowed with consciousness, qualia, experiences, intentions, beliefs, emotions, memories. When we ask whether a machine can think, we're really asking *whether there can be a completely naturalistic account of what a mind is.* I'm a naturalist, so I believe the answer must be yes.

Certainly we're not there yet. Whatever the brain is doing to generate a mind, I doubt that it's only running prespecified algorithms, or doing anything like what present-day computers do. We likely have yet to discover key principles by which a human brain works. I suspect that how and why we think cannot be understood apart from our being alive, so before we understand what a mind is, we'll have to understand more deeply what a living thing is—in physical terms. The construction of an artificial mind probably has to wait until we do.

This understanding will have to address what David Chalmers calls "the hard problem of consciousness": how to account for the presence of qualia in the physical world. We have reason to believe that our sensations of, say, the color red are associated with certain physical processes in our brains, but we're stumped because it seems impossible to explain in physical terms why or how those processes give rise to qualia.

A key step toward solving this hard problem is to situate our description of physics in a relational language. As set out by Leibniz, the patron saint of relationalism, the properties of elementary particles have to do with relationships with other particles. This has been a very successful idea, well realized by general relativity and quantum theory, so let's adopt it.

The second step is to recognize that events or particles may have properties that aren't relational—not described by giving a complete history of the relationships they enjoy. Let's call these *internal properties*.

If an event, or a process, has internal properties, you cannot learn about them by interacting with it or measuring it. If there are internal properties, they aren't describable in terms of position, motion, charges, or forces—i.e., in the vocabulary physicists use to talk of relational properties. You might, however, know about a process's internal properties by being that process.

So let's hypothesize that qualia are internal properties of some brain processes. When observed from the outside, those brain processes can be described in terms of motions, potentials, masses, charges. But they have additional internal properties, which sometimes include qualia.

Qualia must be extreme cases of being purely internal. More complex aspects of mind may turn out to be combinations of relational and internal properties. We know that thoughts and intentions can influence the future.

There's much hard scientific work to do to develop such a naturalistic account of mind—one that's nondualist and not deflationary, in that it doesn't reduce mental properties to the standard physical properties or vice versa. We may want to avoid naïve panpsychism, according to which rocks and wind have qualia. At the same time, we want to remember that if we don't know what it's like to be a bat, we also don't know, really, what a rock is, in the sense that we may know only a subset of its properties—those that are relational.

One troubling aspect of mind from a naturalistic perspective is our impression that we sometimes think novel thoughts and have novel experiences, which have never been thought or experienced before in the history of the world. Little would make sense about the human world of culture and imagination without allowance for the genuinely novel. A century ago, the *Edge* website didn't exist and likely couldn't have been imagined. Yet it exists, and as naturalists we must have a conception of nature that includes it. This must allow novel kinds of things to come to exist in nature.

We're hamstrung by the conviction that nothing truly new can happen in nature because everything is really elementary particles moving in space according to unchanging laws. Without deviating an inch from rigorous naturalism, however, we can begin to imagine how our understanding of nature can be deepened to allow for the truly novel to occur.

First, in quantum physics we admit the possibility of novel properties arising that are shared among several particles in entangled states. In the lab, we can make entangled states of complex systems that are unlikely to have natural precedents. Hence we can, and do, create physical systems with novel properties. (So, by the way, does nature, when natural selection produces novel proteins that catalyze novel reactions.)

Second, Leibniz's principle of the identity of the indiscernible implies that there can be no two distinct events with exactly the same properties. This means that the fundamental events cannot be subject to laws that are both deterministic and simple, for if two events have precisely the same past, their futures must differ. This presumes a physics that can distinguish the future from the past.

Note that quantum physics is inherently nondeterministic.

Does this imply that quantum physics will play a role in a future naturalistic account of mind? It's too soon to tell, and the first efforts in this direction are unconvincing. But what we learn is that a naturalistic account of mind will require deepening our concept of the natural. We can think novel thoughts by which we can alter the future. Novelty must then be intrinsic to how we understand nature, if minds are to be natural. Therefore, to understand how a machine could have a mind, we must deepen our concept of nature.

MACHINES THAT THINK? NUTS!

STUART A. KAUFFMAN
Pioneer of biocomplexity research; affiliate, Institute for Systems Biology, Seattle; author, *Reinventing the Sacred: A New View of Science, Reason, and Religion*

The advent of quantum biology, light-harvesting molecules, bird navigation, perhaps smell, suggests that sticking to classical physics in biology may turn out to be simply stubborn. Now Turing Machines are discrete state (0,1), discrete time (T, T+1) subsets of classical physics. We all know they, like Shannon information, are merely syntactic. Wonderful mathematical results such as Gregory Chaitin's omega—the probability that a program will halt, which is totally noncomputable and nonalgorithmic—tell us that the human mind, as Roger Penrose also argued, cannot be merely algorithmic.

Mathematics is creative. So is the human mind. We understand metaphors ("Tomorrow, and tomorrow, and tomorrow, creeps in this petty pace . . ."), but metaphors aren't even true or false. All art is metaphoric; language started gestural or metaphoric; we live by these, not merely by true/false propositions and the syllogisms they enable. No prestated set of propositions can exhaust the meanings of a metaphor, and if mathematics requires propositions, no mathematics can prove that no prestated set of propositions can exhaust the meanings of a metaphor. Thus the human mind, in Charles Sanders Peirce's "abduction," not induction or deduction, is wildly creative in un-prestatable ways.

The causal closure of classical physics precludes more than an epiphenomenal mind that cannot "act" on the world, be it a Turing Machine or billiard balls, or classical physics' neurons. The current state of the brain suffices to determine the next state of the brain (or computer), so there's nothing for mind to do and no way for mind to do it! We've been frozen in this stalemate since Newton defeated Descartes's *Res cogitans*.

Ontologically, free choice requires that the present could have been different—a counterfactual claim impossible in classical physics but easy if quantum measurement is real and indeterminate. The electron could have been measured to be spin-up or spin-down, so the present could have been different.

A quantum mind, however, seems to obviate responsible free will. False, for given n entangled particles, the measurement of each alters the probabilities, by the Born rule, of the outcomes of the next measurements. In one extreme, these may vary from 100 percent spin-up on the first to 100 percent spin-down on the second, and so on, for n measurements, entirely nonrandom and free if measurement is ontologically indeterminate. If probabilities of n entangled particles vary between less than 100 percent and 0 percent, we get choice, and an argument suggests we can get responsible choice in the Strong Free Will Theorem of John H. Conway and Simon Kochen.

We'll never get to the subjective pole from third-person descriptions. But a single rod can absorb a single photon, so it's conceivable to test whether or not human consciousness can be sufficient for quantum measurement. If we were so persuaded, and if the classical world is at base quantum, then the easy hypothesis is that quantum variables consciously measure and choose, as Penrose and Stuart Hameroff, in "Orch-OR" theory, and others suggest. We may live in a wildly participatory universe. Consciousness and will may be part of its furniture, and Turing Machines cannot, as subsets of classical physics and merely syntactic, make choices where the present could have been different.

THE FUTURE POSSIBILITY-SPACE OF INTELLIGENCE

MELANIE SWAN
Philosopher; science and technology innovator, MS Futures Group; founder, DIYgenomics

Considering machines that think is a nice step forward in the AI debate, as it departs from our human-based concerns and accords machines otherness in a productive way. It causes us to consider the other entity's frame of reference. But even more important, this questioning suggests a large future possibility-space for intelligence. There could be "classic," unenhanced humans; enhanced humans (with nootropics, wearables, brain-computer interfaces); neocortical simulations; uploaded-mind files; corporations as digital abstractions; and many forms of generated AI—deep-learning meshes, neural networks, machine-learning clusters, blockchain-based distributed autonomous organizations, and empathic compassionate machines. We should consider the future world as one of multispecies intelligence.

What we call the human function of "thinking" could be quite different in the variety of possible future implementations of intelligence. The derivation of various species of machine intelligence will necessarily be different from that of humans. In humans, embodiment and emotion have been important elements influencing human thinking. Machines won't have the evolutionary biological legacy of being driven by resource acquisition, status garnering, mate selection, and group acceptance—at least in the same way. Therefore, species of native machine "thinking" could be quite different. Rather than asking if machines can think, it may be more productive to move from the frame of "thinking" that asks "Who thinks how?" to a world of digital intelligences with different backgrounds, different modes of thinking and existence, and different value systems and cultures.

Already, not only are AI systems becoming more capable, but also we're getting a sense of the properties and features of native machine culture and the machine economy—and of what the coexistence of human and machine systems might be like.

Some examples of these parallel systems are in law and personal identity. In law, there are technologically binding contracts and legally binding contracts. They have different enforcement paradigms: inexorably executing parameters in the case of code ("code is law") and discretionary compliance in the case of human-partied contracts. Code contracts are good in that they cannot be breached, but on the other hand they'll execute monolithically even if conditions change.

With regard to personal identity: The technological construct of identity and its social construct are different and have different implied social contracts. The social construct of identity includes the property of imperfect human memory that allows for forgiving, forgetting, redemption, and reinvention. Machine memory, however, is perfect and can act as a continuous witnessing agent, never forgiving or forgetting and always able to re-present even the smallest detail at any time. Technology itself is dual-use, in that it can be deployed for good or evil. Perfect machine memory becomes tyrannizing only when re-imported to static human societal systems, but it needn't be restrictive. This new "fourth-person perspective" could be a boon for human self-monitoring and mental-performance enhancement.

These examples show that machine culture, values, operation, and modes of existence are already different, and this emphasizes the need for ways to interact that facilitate and extend the existence of both parties. The potential future world of intelligence multiplicity means accommodating plurality and building trust. Blockchain technology—a decentralized, distributed, global, permanent, code-based ledger of interaction transactions and smart contracts—is one example of a trust-building system. The system can be used between human parties or interspecies parties, exactly because it's not necessary to know, trust, or understand the other entity, just the code (the language of machines).

Over time, trust can grow through reputation. Blockchain technology could be used to enforce friendly AI and mutually beneficial interspecies interaction. Someday, important transactions (like identity authentication and resource transfer) will be conducted on smart networks that require confirmation by independent consensus mechanisms, such that only bona fide transactions by reputable entities are executed. While perhaps not a full answer to the problem of enforcing friendly AI, decentralized smart networks like blockchains are a system of checks and balances helping to provide a more robust solution to situations of future uncertainty.

Trust-building models for interspecies digital intelligence interaction could include both game-theoretic checks-and-balances systems like blockchains and also, at the higher level, frameworks that put entities on the same plane of shared objectives. This is of higher order than smart contracts and treaties that attempt to enforce morality; a mind-set shift is required. The problem frame of machine and human intelligence should not be one that characterizes relations as friendly or unfriendly but, rather, one that treats all entities equally, putting them on the same ground and value system for the most important shared parameters, like growth. What's most important about thinking for humans and machines is that thinking leads to ideation, progress, and growth.

What we want, for both humans and machines, is the ability to experience, grow, and contribute more, with the two in symbiosis and synthesis. This can be conceived as all entities existing on a spectrum of capacity for individuation (the ability to grow and realize one's full potential). Productive interaction between intelligent species could be fostered by being aligned in the common framework of a capacity spectrum that facilitates their objective of growth, and maybe mutual growth.

What we should think about thinking machines is that we want greater interaction with them, both quantitatively or rationally and qualitatively, in the sense of extending our internal experience of ourselves and reality, moving forward together in the vast future possibility-space of intelligence.

LOVE

TOR NØRRETRANDERS
Science writer; lecturer, Copenhagen; author, *The Generous Man: How Helping Others Is the Sexiest Thing You Can Do*

Making machines that think will be like putting a man on the moon: The effect will be the exact opposite of what everyone expected. The Apollo program didn't launch humanity into a Space Age of cosmic exploration. It led to something much more important: the Earth Age. We left home to explore the universe and discovered for the first time the place we came from. The image of our planet rising on the sky of the moon became the iconic symbol of ecology, fragility, and globalization.

Thinking machines will mean a huge change in the way we understand something much more subtle and alien than machines—ourselves. Teaching machines to think will teach us who we are and how we think. We don't think the way we think we do. Most of what we do in terms of advanced information processing we don't think about at all. We just do it. A child is threatened and we act, immediately. Only afterward do we start thinking about it. A thought appears in our mind—a beautiful, luminescent, and breathtaking thought. It's just there. We didn't think of it before it was a thought.

We aren't consciously aware of most of the information we process when we think. It all happens unconsciously, in our mind, in our body. Right away. We're not even rational in the sense of being logical and explicitly deductive. We're fast, intuitive, and emotional.

Economists believe we are *Homo economicus*—selfish and rational, acting with reason in our own self-interest. But most economic and social interactions deal with fairness, trust, sharing, and long-term relationships. Experimental economics shows us that when we act directly and without hesitation, we're social and cooperative. Only when we start thinking for some seconds do we choose to be selfish.

Unless we deal with computers. When we play economic games with machine counterparts, we tend to be cold and egoistic. You can even measure the difference in our blood flow in the brain and in the hormones in our bloodstream. We think of machines the way economists think about us—as rational, cold-blooded, and selfish. Therefore we treat them as such.

By instinct, we know that humans are more human than when we think of ourselves in the theoretical terms of economics (or other social sciences). We act according to this instinct, but when we think about it we're still under the false impression that we're *Homo economicus*. Building thinking machines will show us that there was a deep evolutionary wisdom in our social instincts: In the long run, it pays much more to be unselfish. It's not truly selfish for you to be selfish, since being unselfish leads to better results for you.

The strategic lesson we'll have to teach machines is all about love.

Robot scientist Hans Moravec has described different biological and technological systems according to their ability to process and store information. At one end: simple, rule-based, and stereotypical creatures, like viruses, worms, and computers. At the other: the truly powerful information processors, like whales, elephants, and human beings. All the creatures with huge capacity are mammals. Their offspring aren't born with the full program for functioning. They go through many years of upbringing before they can act on their own. Their skills aren't specified as rules but as lessons learned from experience. You bang your head on a table until you learn not to. Learning by trial and error. Exploring. This is possible only because the young mammals are taken care of by older mammals. Parenting. Nursing. Love.

Love creates the trust that gives young mammals confidence enough to go out and collect Big Data about the world. And digest it. And heal the wounds.

Love is the recipe for how to grow a human intelligence, a human set of skills, and a human ability to think. To make machines think, we'll have to give them love. It will be more like a kindergarten than a high-tech lab. We'll have to let machines explore all by themselves, do weird things, not just act according to our wants. They'll have to be not tame but wild, acting from their own will.

The challenge is for us to love wild machines that think. We have to get past the ideas of machines that think and of artificial life. Because when something is alive—and therefore able to self-reproduce and to change—it's no longer artificial. When it thinks on its own, it's no longer a machine but a thinking creature. It will be illogical, intuitive, and benevolent. We'll wonder how it became that way. Until we understand that it was created in our own image.

AN UNCANNY THREE-RING TEST FOR *MACHINA SAPIENS*

KAI KRAUSE
Software pioneer; philosopher; author, *A Realtime Literature Explorer*

> in Just-
> spring   when the world is mud-
> luscious the little
> lame balloonman
>
> whistles far   and wee
> and eddieandbill come
> running from marbles and
> piracies and it's
> spring
>
> when the world is puddle-wonderful

That brillig thing of beauty electric touches me deeply as I think about AI. The youthful exuberance of luscious mud puddles, playing with marbles, or pretending to be a pirate, running *weee* . . . all of which is totally beyond explanation to a hypothetical intelligent machine entity.

You could add dozens of cameras and microphones, touch-sensors, and voice output; would you seriously think it will ever go *wee,* as in E. E. Cummings's (sadly abbreviated) 1916 poem?

To me, this is not the simplistic "Machines lack a soul" but a divide between manipulating symbols versus actually grasping their true meaning. Not merely a question of degree or not having got around to defining the semantics yet, but an entire leap out of that system.

Trouble is, we are still discussing AI so often with terms and analogies of the early pioneers. We need to be in the present moment and define things from a new baseline that is truly interested in testing the achievement of "consciousness."

We need a *Three-Ring Test*. What *is* real AI? What is intelligence, anyway? The Stanford-Binet intelligence test, and William Stern's ratio to the physical age as the "intelligence quotient," IQ, are both over a hundred years old! It doesn't fit us now—and it will fit much less with AI. Really it tests only the ability to take such tests, and the ability of truly smart people . . . to avoid taking one.

We use terms like *AI* too easily, as in Hemingway's "All our words from loose using have lost their edge." Kids know it from games—zombies, dragons, soldiers, aliens. If they evade your shots or gang up on you, that is called "AI." Change the heating, lights, lock the garage—we are told that is a "smart home." Of course, these are merely simplistic examples of "expert systems"—look-up tables, rules, case libraries. Maybe they should be labeled, as artist Tom Beddard says, merely "artificial smarts"?

Let's say you talk with cannibals about food, but every one of their sentences revolves around truffled elbows, kneecap dumplings, cock-au-vin, and crème d'earlobe. From their viewpoint, *you* would be just as much outside *their* system and unable to follow their thinking—at least on that specific, narrow topic. The real meaning and the emotional impact their words have, when spoken to one another, would simply be forever missing for you (or requiring rather significant dietary adjustments). Sure, they would grant you the status of a "sentient being," but they'd still laugh at every statement you made, as ringing hollow and untrue—the "Uncannibal Valley," as it were.

It was Sigmund Freud who wrote about "the Uncanny" in a 1919 essay (in a true Freudian slip, he ends up connecting it to female genitalia); then in 1970 Masahiro Mori described "the Uncanny Valley" concept (about the "Vienna hand," an early prosthesis). That eerie feeling that something is not quite right, out of place (Freud's *Unheimliche*). Like a couple kissing passionately—but as you stare at them a little closer, you realize there is a pane of glass between them.

AI can easily *look* like the real thing but still be a million miles away from *being* the real thing—like kissing through a pane of glass: It looks like a kiss but is only a faint shadow of the actual concept.

I concede to AI proponents all of the semantic prowess of Shakespeare, the symbol juggling they do perfectly. Missing is the direct relationship with the ideas the symbols represent. Much of what is certain to come soon would have belonged in the old-school "Strong AI" territory.

Anything that can be approached in an iterative process can and will be achieved, sooner than many think. On this point I reluctantly side with the proponents: exaflops in CPU+GPU performance, 10K resolution immersive VR, personal petabyte databases . . . here in a couple of decades. But it is *not* all "iterative." There's a huge gap between that and the level of conscious understanding that truly deserves to be called Strong, as in "Alive AI."

The big elusive question: Is consciousness an emergent behavior? That is, will sufficient complexity in the hardware bring about that sudden jump to self-awareness, all on its own? Or is there some missing ingredient? This is far from obvious; we lack any data, either way. I personally think that consciousness is incredibly more complex than is currently assumed by the "experts."

A human being is not merely x numbers of axons and synapses, and we have no reason to assume that we can count our flops-per-second in a plain Von Neumann architecture, reach a certain number, and suddenly out pops a thinking machine.

If true consciousness *can* emerge, let's be clear what that could entail. If the machine is truly aware, it will, by definition, develop a "personality." It may be irascible, flirtatious, maybe the ultimate know-it-all, possibly incredibly full of itself.

Would it have doubts or jealousy? Would it instantly spit out the Seventh Brandenburg and then 1,000 more?

Or it suddenly grasps "humor" and finds Dada in all its data, in an endless loop. Or Monty Python's "killer joke"?

Maybe it takes one long look at the state of the world, draws inevitable conclusions, and turns itself off! Interestingly: With a sentient machine, *you* wouldn't be allowed to turn it off—that's murder . . .

The entire scenario of a singular large-scale machine somehow "overtaking" anything at all is laughable. Hollywood ought to be ashamed of itself for continually serving up such simplistic, anthropocentric, and plain dumb contrivances, disregarding basic physics, logic, and common sense.

The real danger, I fear, is much more mundane. Already foreshadowing the ominous truth: AI systems are now licensed to the health industry, Pharma giants, energy multinationals, insurance companies, the military . . . The danger will not come from *Machina sapiens*. It will be . . . quite human.

Ultimately, though, I do want to believe in the human spirit.

To close things off symmetrically with E. E. Cummings: "*Listen: there's a hell of a good universe next door; let's go.*"

FREE FROM US

GEORG DIEZ
Writer, journalist, *Spiegel Online*; former cultural editor,
Süddeutsche Zeitung, Frankfurter Allgemeine Zeitung, and *Die Zeit*

The thinking machine, Turing's turmoil: Does it really change *everything*? It is, after all, a human folly to believe that this is how things work—that a single event can separate time, man, thinking. A self-negating and at the same time self-elevating sentimentality, both optimistic and pessimistic, nihilistic and idealistic.

Because, really, what does it mean? And who's to judge? What is "everything"? And what is "change"? What is before and what is after? We'd first have to agree about the state of affairs, and that itself is difficult enough. Are we free, for example? And free from what? Is biology a system that allows for freedom? To a certain degree, yes. Is democracy a system that allows for freedom? Yes, but only theoretically sometimes, and tragically less and less so. Is capitalism a system that allows for freedom? Not for everybody, that's for sure.

So is freedom, after all, the right approach, the right thing to ask for? Yes, if domination is what we fear from thinking machines. But should this be the way we think about thinking machines? Is negativity equal to critical thinking? Is critical thinking the right way to produce some real insight? Or is this onanistic logic, meant to please ourselves without regard for others and the outside world? Who are we addressing in such a critical way? People we'd like to convince? Is that possible? Or is it a chimera? A strange turn of reason, the conceit of the "enlightened" community?

Not that progress isn't possible. Quite the contrary, and thinking machines speak to this. Maybe the idea of progress itself isn't necessarily tied to the idea of humanity. Maybe humans aren't the eternal carrier of this idea. Maybe the idea will eventually detach itself from humans and develop its own reality. Maybe this is what the thinking machine is all about: a difference, a mirror, a chance to reflect. Free from us. Free from the burden of humanity and history.

Human history is in large part the piling of mythology upon mythology—and then of the strenuous effort to unravel the lot, straighten it out, get it right again. It's as if we'd set up barricades purely in order to remove them, to give us a sense of meaning, of purpose. That was ridiculous, like so much we humans do. So to think about machines means to think about humans less as humans. Which sets us free from all the old lore in which we've been caught up—old concepts of order, life, happiness.

Family, friendship, sex, money—everything could be different. Those aren't the only possible answers to the question of human freedom and how to create it—and more important, how to constrain it. The thinking machine is the necessary question mark behind our existence. It's a blank space, just like everybody's life. It offers the possibility of freeing us from our evolutionary, psychological, neurological assumptions. In a truly antihumanistic humanistic sense, in the romantic tradition of E. T. A. Hoffmann, this could be a poetic and thus a political proposition.

It could free us from *us*.

FLAWLESS AI SEEMS LIKE SCIENCE FICTION

EDUARDO SALCEDO-ALBARÁN
Philosopher; director, Scientific Vortex, Inc.

A "conscious" or "thinking" machine should behave erratically, in a sometimes stupid and sometimes smart way. Human "intelligent" behavior is about unpredictable oscillations between emotions and reason; this is what *Homo sensus sapiens* is about. Paradoxical as it sounds, we call a species characterized as equally and randomly stupid and smart "intelligent."

In first-person perspective, we know we're conscious, although there's no definitive way to prove it. In third-person perspective, it's also impossible to verify that someone or something else is conscious. All we do is perceive signals—sounds, gestures, expressions—and infer consciousness.

Any software or robot can pronounce the words "I am intelligent" or "I am conscious," but those aren't proofs of intelligence. In practical terms, consciousness and intelligence are perceived and attributed. This attribution depends on our empathy and criteria for anthropomorphizing. We tend to infer that others are conscious if they behave, look, or (in Turing terms) answer questions like us. The Turing Test is a social experiment about perceiving and assigning intelligence to a machine, not about proving that the machine thinks. This is a social game we play every day. A "thinking machine" is actually a social machine, not a functional but isolated mind.

The idea of creating a singular intelligent machine that will solve the mysteries of reality through flawless logic and produce a whole new species is in the domain of science fiction. The Singularity idea is not an event horizon but an endless effort.

The human mind is resilient because decentralized networks of other minds and knowledge sustain it. As we grow, we enter those networks via language and concepts that don't accord with perfect logic; we then become resilient minds by navigating and exploring those networks; and finally we leave them as we lose brain capacities—for instance, with Alzheimer's. This is an analogous process: We're never absolutely inside or outside the human-knowledge networks. In this process, words and concepts are characterized by ambiguity. Logic and perfection are present only in artificial languages—mathematics, geometry, software—that we cannot use to communicate in everyday life. Imperfection and ambiguity define human thinking; that's why, even in science fiction, humans usually find unexpected ways to beat the logic of the machines.

Therefore the possibility of a flawless superintelligent machine seems like science fiction: We can never condense the entire knowledge of the world, so we can't teach a machine how to do it. We can teach a machine how to acquire knowledge, but it will always be an unfinished process. This doesn't mean that artificial intelligence is irrelevant. We don't fully understand brains and minds yet, and that makes artificial intelligence and thinking machines more relevant now than ever.

We can solve practical problems by simulating specific elements of the mind through machine learning and deep learning. This is what artificial expert systems currently do. But that doesn't mean we're creating actual minds: Simulating minds is like creating artificial meat that vegans can eat by reorganizing chemical compounds found in plants. The simulated meat tastes like meat, but it's not.

Or we can try to create real meat, not imitate it—for instance by cloning cow cells. Maybe the cloned meat and the replicated mind won't alter society, because we also still have the originals, but they'll take us to a whole new level of understanding. In the end, the efforts to understand—or to simulate or create—minds will be relevant if they improve coexistence. We're well aware that religion, exacerbated ambition, and intolerance can lead to social tragedies because of a failure to achieve the delicate equilibrium between emotion and reason.

As we approach the border between peace and barbarity in various regions of the world, artificial intelligence allows us to integrate all we know and all we need to know for achieving coexistence and balance among the organic machines that we are and (maybe) the inorganic machines that will come.

EMERGENT HYBRID HUMAN/ MACHINE CHIMERAS

MARIA SPIROPULU
Experimental particle physicist, Caltech

So-called artificial intelligence as an emulation of human intelligence is beginning to emerge, based on technology advances and the study of human complexity. The former includes high-performance computing systems tooled with intelligent, agile software—machine learning, deep learning, and the like—and the connection of many such systems in self-organized, autonomous, optimized ways. The latter entails neuroscience, genomics, and new cross-disciplinary fields.

A thinking machine is not, and cannot be, a copy of a human; this is because we cannot yet claim to know the workings of the human brain. But considered as a species developed as the product of advanced human logic, science, and technology, the thinking machine will no doubt surpass the human capacity in many of its functions. With its huge memory and data storage, it will also be able to process all of our knowledge. Tooled impeccably with its data-driven discovery methodologies it will detect unusual patterns in data and learn from them. It will compile everything, surely—but to what end? Human intelligence (hard to define, really) is based on knowledge that produces intuition, hunches, passion—and dares, when it comes down to survival, to conquer new ground and attack the unknown. An almost poetic quest for advancement, innovation, and creativity emerges from the thinking, feeling, dreaming, daring, indomitable, fearless, highly sociable, interacting, independent, and proud human being. Can we code the complex superposition of these attributes to give the thinking machine a fair head start on its evolution from where we stand today? Our own has by now produced an organic, complex intelligence.

There's a lot of technopanic recently regarding machines that think, coming from thoughtful and otherwise fearless human beings. I, for one, am more concerned about humans who are brainwashed, or stop thinking, than about smart thinking machines taking over—mainly because "machine thinking" cannot fully substitute for the full range of human thinking and operation. Even assuming the Cylon sci-fi case—immortal knowledge and consciousness, a sensory system, a powerful memory—the problem remains: Human intelligence (brain, senses, emotions) is complex intelligence. It masters the complex world with tools that connect disparate facts, and it does so efficiently by dropping most information! Even as we prepare machine-learning algorithms and try mimicking the brain with deep neural networks in all scientific domains, we remain puzzled by the mode of connected knowledge, intuition, imagination, and organic reasoning tools the human mind possesses. This is difficult, perhaps impossible, to replicate in a machine. Infinite, unconnected clusters of knowledge will remain sadly useless and dumb. When a machine can remember a fact on its own initiative, spontaneous and untriggered; when it produces and uses an idea not because that idea was in the algorithm of the human who did the programming but because it connects to other facts and ideas beyond the machine's training samples or utility function—then I'll start hoping we can build a totally new branch of artificial species, self-sustainable and endowed with independent thinking.

Meanwhile I foresee the emergence of hybrid human/machine chimeras: human-born beings augmented with new machine abilities that enhance all or most of their human capacities, pleasures, and psychological needs—and to the point where thinking might be rendered irrelevant and, strictly speaking, unnecessary, providing ordinary thinking humans with a better set of the servants they have been looking for in machines.

WHAT IF THEY NEED TO SUFFER?

THOMAS METZINGER

Professor of philosophy, Johannes Gutenberg-Universität Mainz; author, *The Ego Tunnel*

Human thinking is efficient because we suffer so much. High-level cognition is one thing, intrinsic motivation another. Artificial thinking might soon be much more efficient—but will it necessarily be associated with suffering in the same way? Will suffering have to be a part of any postbiotic intelligence worth talking about, or is negative phenomenology just a contingent feature of the way evolution made us? Human beings have fragile bodies, are born into dangerous social environments, and find themselves in a constant uphill battle to deny their own mortality. Our brains continually fight to minimize the likelihood of ugly surprises. We're smart because we hurt, because we can regret, and because of our constant striving to find some viable form of self-deception or symbolic immortality. The question is whether good AI also needs fragile hardware, insecure environments, and an inbuilt conflict with impermanence. Of course at some point there will be thinking machines! But will their own thoughts *matter* to them? Why should they be interested in their thoughts?

I'm strictly against even risking the building of a suffering machine. But just as a thought experiment, how would we go about doing it? Suffering is a phenomenological concept. Only beings with conscious experience can suffer (call this necessary condition no. 1, the C condition). Zombies, and human beings in dreamless deep sleep, coma, or under anesthesia, don't suffer, just as possible persons or unborn human beings who haven't yet come into existence cannot suffer. Robots and other artificial beings can suffer only if they're capable of phenomenal states, if they run under an integrated ontology that includes a window of presence.

Condition no. 2 is the PSM condition: possession of a phenomenal self-model. Why this? The most important phenomenological characteristic of suffering is the sense of ownership, the untranscendable subjective experience that it is *I* who suffers right now, that it's my *own* suffering I'm undergoing. Suffering presupposes self-consciousness. Only those conscious systems that have a PSM are able to suffer, because only they—through a computational process of functionally and representationally integrating certain negative states into their PSM—can appropriate the content of certain inner states at the level of phenomenology.

Conceptually, the essence of suffering lies in the fact that a conscious system is forced to identify with a state of negative valence and cannot break this identification or functionally detach itself from the representational content. Of course, suffering has many different layers and phenomenological aspects, but it's the *phenomenology of identification* that counts. What the system wants to end is experienced as a state of itself, one that limits the system's autonomy because the system cannot effectively distance itself from it. If you understand this point, you also see why the "invention" of conscious suffering via biological evolution was so efficient—and (had the inventor been a person) not only truly innovative but also a nasty and cruel idea.

Clearly the phenomenology of ownership is not sufficient for suffering. We can all easily conceive of self-conscious beings that don't suffer. Suffering entails the NV (negative valence) condition. Suffering is created by the integration of states representing a negative value into the PSM of a given system. Thus negative preferences become negative *subjective* preferences—i.e., the conscious representation that one's own preferences have been (or will be) frustrated. This doesn't mean that our AI system must have a full understanding of what those preferences are. If the system wants not to undergo the current conscious experience again—wants it to end—that suffices.

Note that the phenomenology of suffering has many facets, and that artificial suffering could be very different from human suffering. For example, damage to physical hardware could be represented in internal data formats alien to human brains, generating a subjectively experienced, qualitative profile for bodily pain states that is impossible to emulate, or even vaguely imagine, for biological systems like us. Or the phenomenal character accompanying high-level cognition might transcend human capacities for empathy and understanding—say, with intellectual insight into the frustration of one's preferences or the disrespect of one's creators, or perhaps into the absurdity of one's existence as a self-conscious machine.

And then there's the T (transparency) condition. Transparency is not only a visual metaphor but also a technical concept in philosophy that comes in a number of different uses and flavors. Here I'm concerned with phenomenal transparency, a property that some (but not all) conscious states have and no unconscious state has. The main point is straightforward: Transparent phenomenal states make their content appear irrevocably *real*—as something whose existence you cannot doubt. More precisely, you may be able to have cognitive doubts about its existence, but according to subjective experience this phenomenal content—the *awfulness* of pain, the fact that it is *your own* pain—is not something from which you can distance yourself. The phenomenology of transparency is the phenomenology of direct realism.

Our minimal concept of suffering is thus constituted by four necessary building blocks: the C, PSM, NV, and T conditions. Any system satisfying all four conceptual constraints should be treated as an object of ethical consideration, because we don't know whether they might already constitute the necessary *and sufficient* set of conditions; we're ethically obliged to err on the side of caution. And we need ways to decide whether a given artificial system is currently suffering, or whether it has that capacity or is likely to generate that capacity in the future.

But by definition, any intelligent system—biological, artificial, or postbiotic—that doesn't fulfill at least one of the necessary conditions cannot suffer. Let's look at the four simplest possibilities:

- An unconscious robot cannot suffer.
- A conscious robot without a coherent PSM cannot suffer.
- A self-conscious robot without the ability to produce negatively valenced states cannot suffer.
- A conscious robot without any transparent phenomenal states cannot suffer, because it would lack the phenomenology of ownership and identification.

I've often been asked whether we could make self-conscious machines that are superbly intelligent *and* unable to suffer. Can there be real intelligence without an existential concern?

WILL WE RECOGNIZE IT WHEN IT HAPPENS?

BEATRICE GOLOMB
Professor of medicine, UC San Diego

Many potential paths lead to a technological superintelligence, onto which a supremacy imperative can be affixed—a superintelligence that might enslave or annihilate humankind.

Technology has long outstripped humanity across legion competencies—even many for which evolution designed us. For instance, discerning the sex of human faces is a task we're designed by evolution to do. Yet even there—and already a quarter century ago—computers bested us. This is one among illimitable illustrations that for myriad tasks, computers have long eclipsed humans. Since those primordial days, countless innovations and applications (GPS, drones, deep networks) provide pieces of a puzzle that, when interconnected, offer a profusion of paths toward future extermination or domination of humans by machines. But just as the target for computer "intelligence" shifts as we acclimate to the latest ability, so too the march toward technological supremacy may go unnoticed, as each incremental encroachment is taken for granted.

Must we even await the future? The answer depends how one defines the question.

1. How many steps removed must the human input be, to deem the technology culpable?
2. How clear must the chasm be between machine and human?

3. Must malice aforethought drive humanity's destruction or subjugation?
4. Must *everyone* be killed or enslaved?

Even insisting upon actions far removed from human input, proscribing human/computer fusion (or collusion!), prescribing premeditation and mandating that all humankind be massacred: The potential remains clear. But suppose we relax these constraints?

1. If human input need be at *no* remove: Fretting over whose finger was on the proverbial button (as enshrined in Tom Lehrer lyrics) predates the rise of modern digital technology.
2. Capacity-enhancing wearables/externals (from old-fashioned glasses and ear trumpets to hearing aids, iWatches, and Oscar Pistorius legs) and implantables (cochleas, pacemakers, radio-controlled spinal devices for paralyzed persons) blur the partition between human and machine. The keen and reluctant alike partake, invested with child-finder microchips or adorned with GPS ankle bracelets. Primitive exemplars have long flaunted their destructive potential—explosives belts as wearables, biological-warfare agents (like the smallpox deployed to vanquish Native Americans) as implantables.
3. May humanity's downfall be epiphenomenal? Or must technology "premeditate" human death, decline, or subjugation?

Regarding subjugation: Many now devote their existence to serv(ic)ing technology and nurturing its "evolution." They mine minerals, design devices, craft programs and apps, or abet devices' diaspora—channeling custody to caregivers who can serve and service them—or paying for same. (Humans service technology, enabling technology to better conduct "its" business, even as technology services humans so we can better conduct our own.) A dispassionate onlooker could justly ask which is master and who is slave.

Regarding Death, Decline, Disability: TICs and TIMs (toxic industrial chemicals, toxic industrial materials)—from production, use, distribution, and disposition of technology—and electromagnetic exposures (from technology itself, or communication signals therefrom) contribute to the explosion of oxidative stress (OS—cell injury of a type against which antioxidants protect) and associated human afflictions (cancer, neurodegenerative disease, obesity, metabolic syndrome, autoimmune disease, chronic multisymptom illness, autism spectrum disorder).

The lattermost conditions seem selectively to attack the best and brightest—the would-be superintelligent?—of our own kind, as others also observe. I suggest that since OS injures mitochondria, the energy powerhouses of cells, and since those whose biology disposes them to greater brain connectivity and activity also demand more cell energy, such potentially superpowered persons have a heightened risk of cell damage and death. Shortfalls of energy supply due to OS are magnified in settings of high energy demand. (Even "typical" human brains, at about 2 percent of body weight, consume about 20 percent of the oxygen and 50 percent of the glucose of the total body.) One consequence: The rise of superintelligent computers may already have come at selective cost to the would-be superintelligent among us.

So, yes, in the obvious sense, technology may become superintelligent and elect to annihilate or enslave us. But it may progress to similar ends through less obvious means—and may be in that process as we speak.

METAREPRESENTATION

NOGA ARIKHA

Historian of ideas; coauthor (with Marcello Simonetta),
Napoleon and the Rebel: A Story of Brotherhood, Passion, and Power

The history of humanity and the history of technology are conjoined. We have always used our cognitive capacities to create the objects we needed to survive, from weapons to garments and shelters. The evolution of the human mind is instantiated in the evolution of technology. We've developed a capacity for metarepresentation—awareness of having, and the ability to analyze, our own minds—which is a function of higher-order consciousness. And in order to look at ourselves in the mirror, we've always used technological analogies—compared our minds to the technologies we create. To each era, its machine—from hydraulic pumps to computers.

We have by now created technologies that no single person is able to master. Our creations are starting to escape our own minds. No wonder, then, that we so easily imagine the creations becoming creatures in their own right, endowed with minds as agile as ours, or more agile perhaps. Science fiction imagines perfect robots, indistinguishable from ourselves, embodied, speaking, seemingly feeling, which can fool and even perhaps attack us.

But in thinking conceptually about our own minds, we tend to remain Cartesian dualists. Thinking seems so disembodied an activity that we forget we're not brains in vats, that no amount of microtechnology will re-create the complexities of biology thanks to which our brains function, replete with neurotransmitters, enzymes, and hormones. We are our bodies; we have emotions that are embodied and that deeply inform our thinking processes. Machines are developing task-driven cognitive capacities, but their perfect processing is very different indeed from the imperfect, inconstant, subtle thinking of persons endowed with a sense of self, proprioception, a sense of centeredness, the qualia that distinguish us from "zombies."

Computers excel at processes most of us fumble with, and we increasingly access the world of facts via machines. Much of our memory is assigned to Google, and there's no doubt that our minds are extending more and more beyond our bodies—that we exist within a growing network of disembodied minds and data. Thinking is in part a social capacity; to think is to participate in a collective enterprise, and the complexity of this enterprise is as much a characteristic of the human condition as our embodiment. Machines don't have social lives, anymore than they're embodied in a complex, evolved set of biological tissues. They're good at tasks, and we've become good at using them for our purposes. But until we replicate the embodied emotional being—a feat I don't believe we can achieve—our machines will continue to serve as occasional analogies for thought and to evolve according to our needs.

ENVOI: A SHORT DISTANCE AHEAD—AND PLENTY TO BE DONE

DEMIS HASSABIS
Vice President of Engineering, Google DeepMind; cofounder, DeepMind Technologies
SHANE LEGG
AI researcher; cofounder, DeepMind Technologies
MUSTAFA SULEYMAN
Head of applied AI, Google DeepMind; cofounder, DeepMind Technologies

For years we've been making the case that artificial intelligence, and in particular the field of machine learning, is making rapid progress and is set to make a whole lot more progress. Along with this, we've been standing up for the idea that the safety and ethics of artificial intelligence is an important topic that all of us should be thinking about very seriously. The potential benefits of artificial intelligence will be vast, but like any powerful technology these benefits will depend on the technology being applied with care.

While some researchers have cheered us on since the start of DeepMind, others have been skeptical. However, in recent years the climate for ambitious artificial intelligence research has much improved, no doubt due to a string of stunning successes in the field. Not only have a number of longstanding challenges finally been met but there's a growing sense among the community that the best is yet to come. We see this in our interactions with a wide range of researchers, and it can also be seen from the way in which media articles about artificial intelligence have changed in tone. If you hadn't already noticed, the AI Winter is over and the AI Spring has begun.

As with many trends, some people are a little too optimistic about the rate of progress, going as far as predicting that a solution to human-level artificial intelligence might be just around the corner. It's not. Furthermore, given the negative portrayals of futuristic artificial intelligence in Hollywood, it's perhaps not surprising that doomsday images still appear with some frequency in the media. As Peter Norvig has aptly put it, "The narrative has changed. It has switched from, 'Isn't it terrible that AI is a failure?' to 'Isn't it terrible that AI is a success?'"

The reality is not all that extreme. Yes, this is a wonderful time to be working in artificial intelligence, and like many others, we expect that this will continue for years to come. The world faces a set of increasingly complex, interdependent, and urgent challenges, requiring ever more sophisticated responses. We'd like to think that successful work in artificial intelligence can contribute by augmenting our collective capacity to extract meaningful insight from data and helping us to innovate new technologies and processes to address some of our toughest global challenges.

However, in order to realize this vision, many difficult technical issues remain to be solved, some of which are longstanding and well known in the field. While difficult, these problems can be overcome, but it will take a generation of talented researchers equipped with plentiful computational resources and inspired by insights from machine learning and systems neuroscience. While this is likely to disappoint the most optimistic observers, it will give this community some time to come to grips with the many subtle questions of safety and ethics that will arise. So let's enjoy this new sense of optimism, but let's not lose sight of how much hard work is left to do. As Alan Turing once said, "We can only see a short distance ahead, but we can see plenty there that needs to be done."

NOTES

14 1. Seth Lloyd, *Programming the Universe* (New York: Knopf, 2006).
17 2. Steven Pinker, *The Better Angels of Our Nature* (New York: Viking Penguin, 2011).
60 3. James R. Beniger, *The Control Revolution: Technological and Economic Origins of the Information Society* (Cambridge, MA: Harvard University Press, 1986).
125 4. Nick Bostrom, *Superintelligence: Paths, Dangers, Strategies* (New York: Oxford University Press, 2014).
153 5. G. K. Chesterton, *Heretics* (New York: John Lane, 1905).
178 6. I. J. Good, "Speculation Concerning the First Ultraintelligent Machine," *Advances in Computers*, vol. 6, 1965.
182 7. Kevin Kelly, "The Technium," *Edge*, entry February 3, 2014, https://edge.org/conversation/the-technium [accessed July 21, 2015].
223 8. Alan Turing, "On Computable Numbers, with an Application to the *Entscheidungsproblem*," *Proc. Lond. Math. Soc.* 42, series 2 (1936–7): 230–65.
238 9. Steven Pinker, comment on "The Myth of AI," *Edge*, entry November 14, 2014, http://edge.org/conversation/the-myth-of-ai#25987 [accessed July 21, 2015].
370 10. Hannah Arendt, *The Life of the Mind*, vol. 1 (New York: Harcourt Brace, 1978).
388 11. Nick Bostrom, "Existential Risks," *Jour. Evol. & Technol.* 9, no. 1 (2002), at http://www.nickbostrom.com/existential/risks.html.
468 12. Doris Jonas and David Jonas, *Other Senses, Other Worlds* (New York: Stein & Day, 1976).
486 13. David Romer, "Do Firms Maximize? Evidence from Professional Football," *Jour. Political Econ.* 114, no. 2 (2006): 340–65.
493 14. Karen A. Frenkel, "The Wisdom of the Hive: Is the Web a Threat to Creativity and Cultural Values? One Cyber Pioneer Thinks So," *Sci. Amer.*, February 16, 2010.

BOOKS BY JOHN BROCKMAN

WHAT TO THINK ABOUT MACHINES THAT THINK
Available in Paperback and eBook

John Brockman has harnessed the brainpower of some of the world's greatest scientists, artists, and philosophers to answer one of today's most consequential questions: what to think about machines that think.

THIS IDEA MUST DIE
Scientific Theories That Are Blocking Progress
Available in Paperback and eBook

Profound, engaging, thoughtful, and groundbreaking, *This Idea Must Die* will change your perceptions and understanding of our world today . . . and tomorrow.

WHAT SHOULD WE BE WORRIED ABOUT?
Real Scenarios That Keep Scientists Up at Night
Available in Paperback and eBook

Drawing from the horizons of science, today's leading thinkers reveal the hidden threats nobody is talking about—and expose the false fears everyone else is distracted by.

WHAT IS YOUR DANGEROUS IDEA?
Today's Leading Thinkers on the Unthinkable
Available in Paperback and eBook

The world's leading scientific thinkers explore bold, remarkable, perilous ideas that could change our lives—for better or for worse.

THIS WILL MAKE YOU SMARTER
New Scientific Concepts to Improve Your Thinking
Available in Paperback and eBook

"A winning combination of good writers, good science and serious broader concerns."
—*Kirkus Reviews* (starred review)

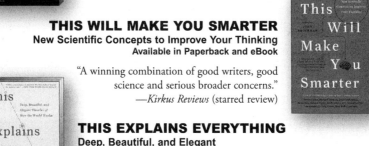

THIS EXPLAINS EVERYTHING
Deep, Beautiful, and Elegant Theories of How the World Works
Available in Paperback and eBook

"A smorgasbord of ideas." —*Kirkus Reviews*

Also Available in Paperback and eBook

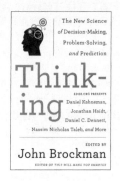

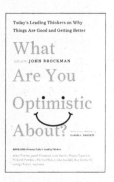

Discover great authors, exclusive offers, and more at hc.com.

Available wherever books are sold.